T0291120

DEEPWATER ALCHEMY

DEEPWATER ALCHEMY

Extractive Mediation and the Taming of the Seafloor

Lisa Yin Han

UNIVERSITY OF MINNESOTA PRESS

MINNEAPOLIS | LONDON

The University of Minnesota Press gratefully acknowledges the financial assistance provided for the publication of this book by the Humanities Institute at Arizona State University.

A different version of chapter 1 was previously published as "The Blue Frontier: Temporalities of Extraction and Salvage at the Seabed," in "Science Studies and the Blue Humanities," special issue, *Configurations* 27, no. 4 (Fall 2019): 463–82. Chapter 2 was previously published as "Sonic Pipelines at the Seafloor," in *Media + Environment* 3, no. 2 (May 18, 2021): 1–23, https://doi.org/10.1525/001c.21392.

Published by the University of Minnesota Press
111 Third Avenue South, Suite 290
Minneapolis, MN 55401–2520
http://www.upress.umn.edu

ISBN 978-1-5179-1593-3 (hc)
ISBN 978-1-5179-1594-0 (pb)

A Cataloging-in-Publication record for this book is available from the Library of Congress.

CONTENTS

INTRODUCTION
Making Sea into Land

The sediments are a sort of epic poem of the earth. When we
are wise enough, perhaps we can read in them all of past history.
—Rachel Carson, *The Sea Around Us*

I grew up in Boulder, Colorado, amid landlocked trees, plains, and the
Rocky Mountains. My childhood friends know me as a poor swimmer.
I always preferred to have my feet on solid ground. As a kid, however,
I latched on to stories of the salty seas and let my imagination swim
wild, dreaming of a day when I could move to the coast. There was one
Chinese myth in particular that my parents used to read to me, and as I
began writing this book, gazing over the bluffs by UC Santa Barbara,
the story came back to me. It goes something like this.

A young girl named Nü Wa is playfully swimming in the Eastern Sea,
when she ventures too far out from the shore and accidentally drowns.
Upon her death, the girl's soul is transformed into a bird called Jing Wei.
Feeling despair and resentment over the abrupt end to her previous life,
Jing Wei vows to fill up the sea in order to prevent others from encoun-
tering the same fate. Day in and day out for eternity, she flies in materials
from the Western Mountains and drops them into the vast ocean—stick
by stick, stone by stone.

I loved this story for the way the depths of the open sea seemed to
reflect a depth of human feeling: danger, sorrow, alienation, injustice,
determination, and a never-ending tedium. In this myth, the ocean is a
powerful agent, callous and indifferent to the life it swallows. The conflict
between Jing Wei and the deep ocean paints a portrait of humanity as a
species living in a world that will always exceed us; the struggle between

1

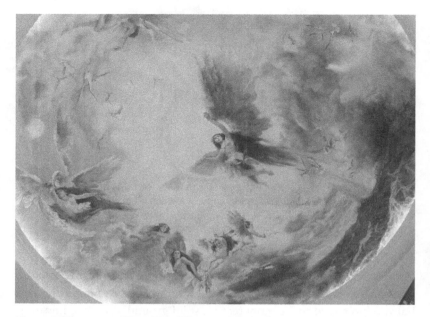

Figure 1. The story of Jing Wei filling up the sea depicted on the ceiling mural of Tianjin Railway Station. Artists: Qin Zheng, Wang Yuqi, Wu Enhai, et al. Photograph by Wuyouyuan, 1988.

humans and nature, earth and water, reflects a primordial balance between the elements. In many ways, stories from the West are similar, placing monsters in the depths and casting the ocean as a dual source of fear and conquest. Poseidon, the Kraken, Atlantis, the Leviathan—all around the world, humans have imagined the sea as an entity that feeds us, connects us to each other, and calls to us to risk our lives.

The part of Jing Wei's story that churns my imagination the most, however, is Jing Wei's desire to make sea into land. The more I have talked to the people who spend their lives thinking about the deep ocean, the more I have come to see this fantasy of transforming fluid into solid as a recurring refrain in marine research today. There are instances where the Jing Wei story becomes literal, such as with the building of artificial islands in the South China Sea.[1]

Other plans are more far-fetched. In the 1960s, optimism about the development of new deep-sea salvage technologies led public figures like

Senator Claiborne Pell to make wild speculations about how the sea-floor could be used by man: "Storage of machinery in the future may be feasible in the ocean at depths below 5500 feet. . . . Wouldn't it be handy if we could tow our surplus warships out to sea, open their sea cocks, and stack them on the bottom until needed?"[2] This premise seems silly today with our knowledge of deep-sea pressures and the corrosive action of seawater. Yet, an influx of new underwater technologies has only exacerbated this tendency to project terrestrial utopias onto the seafloor. From seafloor museums to entire seafloor civilizations, remote sensing and videography enabled by deep-submergence vehicles, underwater cables, and satellites opened up new ideas about the deep sea as a potentially functional and stable space for human existence. In Jing Wei's world, perhaps, sticks and stones were the media through which this transformation from sea to land was performed. Today, we venture to the deep ocean to fill it up with media technologies: camera by camera, sensor by sensor, platform by platform.

Our efforts to visualize and extract from the seabed are challenged by the vast environment of the deep sea, the most unknown region on Earth. Occupying over 60 percent of the planet, deep water typically refers to between 1,000 and 5,000 meters of depth—the bathypelagic, abyssopelagic, and hadopelagic zones. Ecologists also refer to this as the benthic zone—the environment at the bottom of a body of water. The abyss brings with it a distinct cultural imaginary. Ann Elias explains that all the world's oceans were once imagined to be bottomless, which made early ocean films by figures like John Ernest Williamson pleasurable for the thrilling idea that a person could go somewhere and never experience an end.[3] Over time, however, the ocean has been partitioned legally, scientifically, and figuratively. Shallow waters have come to reassure and entertain us with their tropical reefs and vibrant colors, promising, as Elias contends, "that the eye will be able to make sense of every object."[4] By contrast, the depths have been associated with mystery, unknowability, and darkness. Over the past century, as humans have continued their endeavors to learn about the planet's abyss, they have found new ways of making the deep sea accessible to the human senses, just as the shallows already are. From trawlers, to sounding lines, to high-definition underwater cameras, technology has enabled us to discover new species and reveal secrets about our deep ocean like never before.

These revelations are what have conditioned the possibility for a new deepwater alchemy.

The title of this book refers to alchemy in more than just a figurative sense. Historically speaking, alchemy was a practice focused on transmutations of substances for profit through a systematic application of elemental knowledge. Headed by figures like Thomas Norton, Roger Bacon, and Johann Joachim Becher, it was once a respectable medieval science that flourished in the West during the Renaissance era, from the fourteenth to sixteenth centuries. Stretching even further back, historian William Newman argues the development of alchemical science during late antique Greco-Roman Egypt led to a shift from mimesis of nature to a naturecultural relationship in which "alchemists began to view the products of their workshops as no different from the natural exemplars upon which they based their designs."[5] While ostensibly this eliminated an ontological distinction between the artificial and the natural, in practice it reified the notion of human conquest over nature. This is not, however, a version of humanity that acts externally on nature. Within the framework of alchemistic science, human exceptionalism emerges from a positioning of nature and culture as selfsame. This is the sense of alchemy I hope to evoke. For my purposes, alchemy, the original science of essences, is a helpful way of describing the material-discursive becoming of the deep ocean, where power is encoded into naturalizations of human technological feats.

Ocean mediation and engineering has long been positioned as a mechanical conquest over nature, achieved through studies and transformations of minerals, fluids, and other kinds of matter. In order to create sound images of the seafloor for deep-sea drilling, we had to study and understand the interactions of sound in seawater. In order to create in situ sensors, submersibles, and platforms, we had to understand how various materials corrode in water and the extent to which certain materials—natural and artificial—could repel living organisms. In order to use sensing instruments like nephelometers, we had to understand the principle behind how light scatters over sediments. And to find and predict the location of hydrocarbons and mineral resources, we had to study how sediments settle. Indeed, I am not the first to describe oceanic mediation as a form of alchemy. For instance, Deborah James describes a corpus of photographs of the Carysfort Reef as constituting a "speculative alchemy" in the way it archived changing human relationships to nature and centered the need for human interventions.[6] While this example foregrounded an environmentalist perspective on reefs, a similar speculative alchemy animates the call for deep-sea extractions.

When it comes to the seabed, the true alchemy, perhaps, is the continued naturalization of an association between surplus capital and power. Historian Pamela Smith explains that during the Holy Roman Empire, alchemy provided a language for commerce, capturing the relationship between artisans and their labor in a way that foregrounded material increase.[7] Today, ocean engineering occupies a similar role in helping to produce the seafloor as a resource frontier. Our reasons for exploring the deep are premised on our perception of the oceans as fertile. This is complemented by regulations that stress the common governance of the ocean's resources. As Katherine Sammler writes, "Transforming the seabed from *aqua incognita* to productive metallurgist necessitated new understandings of sovereignty over the seas and re(in)scriptions of geopolitical land/sea territorial boundaries." This aim was in part accomplished by the 1982 Law of the Sea, which "captured the oceans in a vision of freely flowing commodities and properly fixed resources."[8]

Like outer space, the deep ocean is a region in which no nations are supposed to exert sovereignty. This concept of a maritime commons has its origins in the concept of the "freedom of the seas," or *mare liberum*, proposed by Hugo Grotius in 1609, and was accepted until the nineteenth century, when coastal state demands for customs zones, exclusive fishing rights, and mineral resource exploitation caused conflicts with existing law. In the wake of these disputes, the Third United Nations Conference on the Law of the Sea (1973–82) sought to standardize territorial boundaries and modernize regulation of nonterritorial waters, specifically addressing concerns about offshore resource exploitation, including oil, minerals, and fish.[9] Among the outcomes of this international conversation was the creation of exclusive economic zones, which extended sovereign rights to exploit underwater resources to two hundred nautical miles from a coastline.

Given these regulatory and technological developments, the sea comes to resemble land in two ways: first, through the achievement of easy visual access, and second, through the extension of extractive capitalism, wherein the extraction of riches from the Earth is the primary means by which contemporary society sustains itself.[10] On this second value, the sea becomes a speculative replacement for land. As more and more of our resources above the water are depleted, our futures are unquestionably tied to the fate of the deep ocean.[11] Offshore oil and gas exploration continues to expand, while the automobile industry, in addition to "green" technologies like wind turbines and solar panels, is driving a demand for mineral deposits found on the ocean floor. The result is that our technical ability to explore and visualize the ocean depths is better than ever, contributing to further extractions in the oceans. Knowledge about elemental transmutation is the basis of underwater mediation. And underwater mediation, in turn, enables the final step of alchemy: the transformation of the elemental into the economic.

Blue Economy

By studying nature, oceanographers and ocean engineers fabricate a new natural order that serves both to fix oceanic space as an extractive frontier and to accommodate the material impacts of human mediation on the ocean bottom. This new order comes with a name, and that name is *blue economy*.

A few years ago, I was seated in the back of an auditorium where hundreds of oceanographers, marine scientists, engineers, and industry researchers attended the welcoming event of OceanObs'19, the decadal conference for ocean observers around the world. As speakers rotated in and out of the podium on the main stage, a sense of the attendees' primary commitments started coming into focus. On the one hand, the oceans are in trouble, and no one who works in this field is debating the ecological destruction wrought by anthropogenic climate change. The effects of this global environmental crisis—from sea-level rise, to ocean acidification, to mass extinction—are more pronounced and happening faster in our oceans than in many other ecosystems on the planet. But there were other concerns as well. Buzzwords like "blue economy," "ecosystem services," and "natural capital" hinted at a parallel concern of this conference, one that was not separate from the environmental problems but rather seemed to wrap itself around them.

I first heard a definition of "blue economy" as it was elucidated at OceanObs'19 by Peer Fietzek, the manager of science and research at

KM Contros, a German company specializing in developing and marketing underwater sensors: "A blue economy is one which strikes the right balance between reaping the economic potential of our oceans with the need to safeguard their longer health." Phrases like this and their explicit ties to the language of capitalism seem to encode a rising anxiety about maintaining current trajectories of growth amid strained ecosystems, and a financially dependent relationship between ocean observers and extractive industries—offshore drilling, mining, fishing, and the like. Situated among like-minded peers, these concerns were often made explicit during the Q&A discussions. One participant asked, "How do we change the perception that economic development and environmentalism are at odds?" Another asked, "How can we balance the short-term return of investments desired by politicians/policymakers and the real/perceived longer-term return of ocean observations?" At first blush, these questions express a yearning for cohesion, perhaps even compromise. Yet as one marine scientist responded, the idea that capitalism and environmentalism are at odds "isn't just a perception—it's a fact. But it's not about supporting all the industries, it's about changing the industries to become more sustainable." Even as blue economy rhetoric implies symbiosis, in many ways the phrase only draws more attention to the fundamental contradictions between industry and sustainability.

Some have referred to the ideology behind blue economy as "green capitalism": the belief that the protection of finite planetary resources is somehow reconcilable with a capitalist model of infinite growth.[12] This is indicative of a particular kind of Eurocentric assumption that investments in certain "green" technologies (and the destructive mining and disposal activities that accompany them) will enable durable economic growth. Environmental humanists have often critiqued green capitalism as a form of greenwashing (or in this case bluewashing) for the way it ultimately justifies the continuation of extraction as a necessary evil. Activities such as seabed mining and offshore drilling are also excused due to the ocean's remoteness from denser human settlements; advocates often say that it is better to destroy the seabed than to destroy terrestrial spaces in closer proximity to human life. I contend, however, that this is a false choice. It assumes a version of the ocean as purely a resource frontier, ignoring both the ecological connections between sea and land and the many vital, rich, and long-standing relationships between humans and the oceans that include sustenance, recreation, spiritual connection, and an ethics of responsibility toward nonhuman others.

Extraction has its own enchantments. Scholars of religion have argued that the worship of green growth resembles providentialism in the way it sees extraction as inevitable and views the market as an uncompromising force that subsumes all human agency.[13] Indeed, my on-the-ground observation and interviews largely confirm that this perspective still reigns. Neoliberal faith in the market is buttressed by popular and scientific ocean mediations that further brand the ocean as a space for exploitation.[14] Put simply, blue economy is a story we tell ourselves about the ocean. My task in this book is to unpack how that story is told through data, through technology, and ultimately through oceanographic mediations.

The field of critical media studies is uniquely suited to examine the cultural and technological formations around ocean observation and extraction. A core tenet of the discipline is the insistence that what matters is not just what we tell each other but also how we tell it. That includes the choices we make about what to include and what to exclude, the strategies we use to order and disseminate information, and the technologies we rely on to capture information and bring our ideas to life. The seafloor has yet to be discussed explicitly within a theory of mediation that focuses on the cultural production of knowledge and the uneven distribution of risks and benefits by ocean industries. Yet all knowledge about the deep ocean is contingent on the use of locative, datalogical, and representational media technologies. What I hope to do is center the role of ocean industries in the production of seafloor media and thereby attend to emerging relationships between humans, media technology, and benthic environments.

Recognizing the critical importance of an ocean-centric perspective on modern technological society, my book starts with the physical impacts of media technologies and infrastructures on ocean ecologies. I then explore the technicity of seafloor mediation as it relates to imperatives to exploit mineral and cultural resources at the ocean bottom. I ask, how have global powers historically imagined ocean wilderness, and how do these ideologies influence the production, use, and regulation of underwater media technologies? How are seafloor data interpreted within existing social structures? And flipping this, what is the ecological impact of mediation on ocean environments? What can a media theorist learn from thinking about our sociotechnological relationships to ocean environments prior to and during acts of media representation? Answering these questions requires not only attending to the materiality of the seafloor but also unpacking the ethical and ideological negotiations between scientists, environmentalists, and industry actors in the midst of climate change and growing offshore resource extractions. In this book I will peek behind the curtain of rhetoric that ocean scientists, engineers, and stakeholders use with each other, and I hope in the process to open an aperture into richer, more expansive tellings about what we do at sea and why it matters. Ultimately, I want my readers to come away with an understanding of the ocean's future that moves beyond utilitarian nature, beyond fears of resources running out, and importantly, beyond the ties of capital that bond this community to a corporate tunnel vision of blue economy, technological determinism, and extraction.

Representing and Mediating the Oceans

According to Raymond Williams, historical development of the concept of *mediation* focuses on two senses of the term: the idea of an intermediary between two poles (the medium as conciliatory), and that of form, "an activity that directly expresses otherwise unexpressed relations."[15] The first representational perspective on mediation, which derives largely from film studies, tends to collapse mediation with communication, emphasizing the medium's role in expression. However, representational studies of media can also refer to fidelity and communication in relation to broader social regimes. For instance, cinema scholar Jonathan Beller argues that cinematic visuality transformed the assembly line, articulating capital expansion to consciousness, and political economy to a psycho-symbolic order: "As in the history of factory production, in the movie theatre we make and remake the world and ourselves along with it."[16] As Beller would have it, cinema was an apparatus that gave voice to techno-capitalist relations, thereby extending the social logics and cultures of productivity. Applying this lesson, we can think about how contemporary media formats play a role in representing and articulating existing logics of productivity and political economy when it comes to the global ocean. The ability to predict, traverse, and mediate seawater has always been wrapped up in questions of geopolitical power, empire, and economic exchange. And power, as Stuart Hall has contended, "can never be bracketed out from the question of representation."[17]

My approach to unpacking these representational power dynamics relies on the cultivation of place-based oceanic knowledge. Scholars such as Steve Mentz and Stacy Alaimo have seized on the term "blue humanities" to describe this kind of critical cultural studies work that shifts attention away from land toward water.[18] By submerging critical perspectives into ocean environments, blue humanities scholars aim to change the way the ocean is positioned hegemonically as a mere backdrop for human action, instead animating it as a constitutive part of human culture and history. Within this neat moniker of "blue humanities," however, are disciplinary fractures and diverse values that demand reflection, as they are easily erased by efforts to rebrand or unify humanities scholarship within the neoliberal academy. Media representation of the oceans embeds ideologies that stem largely from Eurocentric frameworks for knowledge, and this includes early blue humanities work, which was largely centered within Western Anglophone literary traditions. Acknowledging this, it becomes all the more important to validate decolonial and feminist perspectives which remind us that human systems require diverse modes of thought. It is telling to me that many researchers working on indigenous ocean knowledges, water rights, ocean militarization, and environmental justice have refused the title of blue humanities, opting instead for terms such as "critical ocean studies" or "hydro-humanities."[19] Elizabeth DeLoughrey argues that part of the problem of the blue humanities is its complicity in reifying a "material and intellectual (blue) spatial fix" that produces the oceanic body as a space of capitalist and creative production.[20] Having waded through oceanographic waters, I have come to share her critique, as I have seen firsthand this use of "blue" to "bluewash" extractive activities, much in the same way "green" has already been co-opted to signal sustainability among corporate entities. Nevertheless, I think that within each of these formations is a shared desire to think seriously and compassionately about human relationships to aquatic spaces, and for that reason I find I am not driven to pick sides. You may call this book a work of the blue humanities, of critical ocean studies, or the hydro-humanities—all three names work for me.

Beyond a focus on representing oceans, my media theoretical inter-ventions are also framed by nonrepresentational, new materialist ontolo-gies of media, especially as they manifest in fields like media infrastructure studies and elemental media studies. Indeed, I see infrastructure studies as environmental in the way it brings in a consideration of how land-scapes and topographies are mediated. Infrastructural studies of media like the telegraph or the fiber-optic cable focus on these material con-straints of meaning-making, looking toward both physical arrangements and structures for knowledge that intervene across both space and time. Feats of engineering such as the building of hardy ship hulls or the pro-duction of seafloor infrastructure are hugely influential in how we inter-pret and imagine our ecological conditions. Following this intellectual lineage, I am interested in both "hard" and "soft" infrastructures—the "stuff you can kick," as Lisa Parks puts it, and the knowledge practices, user networks, and conventions that determine how underwater tech-nologies are designed, deployed, and interpreted.[21]

Our imaginaries of the ocean are often premised on the erasure of the infrastructures and interfaces that filter and arrange information about the sea for our consumption. Scholars such as Parks and Rafico Ruiz have contended that infrastructures are often purposely made to be invisible, even as they play a direct role in manifesting and embed-ding cultural ideologies into lifeworlds. This invisibility is even more of a challenge given the remoteness of the seafloor from human settle-ments. Nevertheless, deep-sea infrastructures and their environments support much of our technological society, and as media scholar Nicole Starosielski demonstrates, it is clear that ocean industries and state actors themselves understand that breakdowns in these infrastructures would have dramatic effects on the everyday lives of people all over the world.[22]

Infrastructural analyses also allow us to think about mediation and media technology beyond their intended uses. I think of Lisa Gitelman's challenge to "hold electronic networks from the wrong end," a consider-ation of how infrastructures are built on top of one another in sometimes unexpected ways, like stapling flyers on telephone poles.[23] Underwater media platforms and infrastructures likewise have lives outside their func-tion as data gatherers. Another generative area for these unexpected infrastructural emergences is in the area of waste. For instance, Steven Jackson has shed light on acts of repair and the afterlives of junk.[24] Media infrastructural perspectives are similarly apparent in the works of Jennifer

Gabrys, Jussi Parikka, Richard Maxwell, and Toby Miller, who attend to the entire production chain of media technology, from the use of raw materials to the environmental impacts of e-waste.[25]

Thinkers across environmental and elemental media studies share this commitment to examining mediation beyond the instrumentality of delivering messages or images. Extending the work of new materialist thinkers such as Karen Barad, Jane Bennett, Manuel DeLanda, and others, elemental media theorists such as John Durham Peters, Melody Jue, and Yuriko Furuhata have chosen to widen the conceptual breadth of the word *media* to encompass the physical environments and large-scale infrastructures that make mediation possible. Many see environments themselves as infrastructural media. "If media are vehicles that carry and communicate meaning," Peters writes, "then media theory needs to take nature, the background to all possible meaning, seriously."[26] In our oceans, light is refracted differently than in air, and devices like underwater cameras must take into account this refraction as well as variables like pressure and salinity in their very design. Seawater and pressure are thus part of the mediating apparatus. Recognizing this, thinkers such as Jue advocate for conceptual displacements of media theory, submerging existing media concepts like "inscription" into materially situated contexts such as the ocean.[27] Jue's tactics of conceptual displacement and submersion in particular are central to this book's methodology. In fact, my primary theoretical formulation of extractive mediation in the next section depends on these infrastructural and elemental perspectives on mediation, which emphasize the ways in which technological mediation adapts around material environmental conditions.

While it is not limitless, elemental mediation presumes mediation as open and entangled rather than bounded, enfolding ecological relationships as well as material transformations into the concept of media. For some, this openness reduces the efficacy of the "media" concept altogether, leading some scholars to ditch it for alternative frameworks like "cultural techniques" or, more simply, "mediation."[28] For instance, media theorists Sarah Kember and Joanna Zylinska propose a shift from "media" to "mediation," which stresses a process of emergence, or becoming with the technological world. This includes "the acts and processes of temporarily stabilizing the world into media, agents, relations, and networks."[29] By replacing discrete "media" objects with a process of "mediation," the authors maintain part of the core premise of mediation as a formal activity that expresses relations. However, they also imply that processes of mediation precede and exceed media, agents, relations, and networks; they include relationships between humans and the resources, infrastructures, and environmental conditions that enable the movement and dissemination of, perceptual attunement to, and the creation of meaning.

One aspect that draws me to elemental and process-oriented theories is the way they deconstruct the assumption that mediation is applied by someone onto something from a distance, creating a third realm between sender and receiver. That is to say, mediation can also happen before any kind of differentiation or distancing between "sender" and "receiver," the way skin mediates sunlight, or the way agar mediates biological cultures. You could describe this more immediate, experiential sense of mediation as "radical mediation," as Richard Grusin does. Grusin posits that radical mediation includes experiences of "that which is felt, embodied, near—not distant from us, and thus not illuminated or pictured, but experienced by us living, embodied human and nonhuman creatures."[30] This way of thinking about mediation fundamentally problematizes the idea of an objective observer, which is an intervention that resonates particularly well with feminist and decolonial perspectives. In fact, there is a long tradition of this type of questioning in feminist science studies. Donna Haraway, for instance, famously advocated for partial and situated knowledges in response to the dynamism of computational meaning-making and its tendency to occupy positions of objectivity and hegemonic authority.[31]

Additionally, studies of the elements often include nonvisual perspectives on mediation as a way of attending to questions of immediacy and nonobjectivity. A multisensory perspective on mediation is especially important for the deep sea. Given that the twilight zone in particular is not easily penetrated by light, underwater mediation has never been limited to visual interfaces. In chapter 2, for instance, I explore sonic mediations using a vibrational ontology of sound, which exceeds a human perceptual framework for listening at a distance. Other ocean sensors account for light (on both visible and invisible wavelengths), pressure, chemicals, and heat. Ocean mediations, by and large, subscribe to multisensory regimes of knowledge. Like the seabed itself, the media interfaces we create for the benthic ocean are volumetric and layered like bedding.

Finally, ocean representation enfolds more than just humans. While human networks and decision making undoubtedly affect the health of our oceans, fields such as science and technology studies (STS) and the environmental humanities have stressed the importance of validating nonhuman agencies. In their foundational 2010 article in *Cultural Anthropology*, S. Eben Kirksey and Stefan Helmreich define multispecies ethnography as an approach that "centers on how a multitude of organisms' livelihoods shape and are shaped by political, economic, and cultural forces."[32] Ursula Heise builds this multispecies ethics further in her proposal for "multispecies justice," which ties multispecies perspectives to differential experiences of risk from resource exploitation and extraction. As she explains, multispecies justice "will need to be accountable not just to the ontological differences between species, but also to the cultural differences in divergent understandings of justice."[33] For me, the ethical core of this book necessitates attending to and representing the experiences of nonhuman life, along with the frictions and affiliations that emerge through multispecies encounter. In line with both multispecies justice and multispecies ethnography, multispecies perspectives are included throughout my chapters, from the plight of hydrothermal vent shrimp to the perspectives of deep-diving whales. Building on Anna Tsing's mode of portraying a biodiverse natural-social landscape, I aim to "give the nonhuman species as much liveliness as the humans."[34]

Extractive Mediation and Dark Mediation

Thus far I have introduced terms such as *deepwater alchemy* and *blue economy* to describe a corporate vision for deep-sea development. Next, I will introduce a conceptual framework for the media work that enables this capitalist enclosure of the seas. My focus on mediation is intended to address what I see as a major gap in how engineers and marine scientists envision the work of underwater mediation. Seafloor mediations often assume highly anthropocentric perspectives that presume an ontological separation between processes of representation and the "real" resources that are represented. In fact, industries like offshore drilling rely on the presumption that representational media are noninvasive and that extraction "starts" with material displacement. My framework of extractive mediation counters this perspective by putting acts of material conversion in conversation with sensing, imaging, data visualization, and modeling activities. Rather than focus on individual acts, we can examine these activities in relation to one another, attending to the vital exchanges that both condition and exceed the terms of resource conversion.

As I define it, extractive mediation is a condition wherein processes of mediation participate in nonreciprocal acts of removal, accumulation, and domination. We can understand how extractive mediation comes about by first examining extractivism itself. For me, extraction is simultaneously a relation, a narrative, and a situated activity. Naomi Klein defines extractivism as a "nonreciprocal, dominance-based relationship with the earth, one purely of taking." This anthropocentric and utilitarian perspective on nature leads to "the reduction of life into objects for the use of others, giving them no integrity or value of their own."[35] In his work on indigenous media, Jordan Kinder likewise emphasizes the idea of nonreciprocity as a key ingredient of extractivism and thus offers reciprocity as its remedy.[36] Extraction can articulate a human-nonhuman relation. But it is also its own form of world-making, altering environments based on a story of endless accumulation. This world-building aspect of extraction includes our existing ideas about the role of technology in environments.

Extractive narratives are animated in extractive zones, areas that, as Macarena Gómez-Barris explains, are marked as biodiverse and actively reduced and converted into capital.[37] Extractive mediations include the infrastructures and media technologies that articulate and thereby produce a paradigm of extractivism for that resource frontier. Not all environmental mediations are extractive, but all extractions imply mediation —they are activities that stabilize and instantiate a particular worldview along with its affects, experiences, and (nonreciprocal) relations. The extractive mediations I describe in this book are seldom cinematic, but they are a form of fantasy. Extractive mediations often include prospecting, a type of looking ahead that constructs the landscape as itself a prospect. Mining the etymology of this term, Lowell Duckert muses, "The *prospect* could be a view that faces a mountain, a consideration of something in the future to be gained from viewing it, and the action of facing it. . . . *A prospect* is *a* look and *to prospect* is *to* look forward across space and time—high and low, past and future—and sometimes all at once."[38] Brian Jacobson delves into prospecting further as it pertains to the corporate filmmaking by petroleum companies, arguing that "prospecting defines the oil industry's inexorable urge to detect, visualize, and interpret subterranean carbon worlds. . . . To succeed in oil, one must look ahead."[39] Prospecting is a technique that imagines the seabed as already resource—as first and foremost a reservoir of hydrocarbons and glittering minerals and tasty fish and lost treasures.

Extractive mediations do not only create prospects, however. Historically, they have also denied prospects to certain peoples and ecologies, and this return to colonial logics is also what makes certain mediations extractive. As Frantz Fanon conceives it, colonization is a project that robs the colonized of temporal belonging—of their historical pasts, and of their futures: "Ideally, the present will always contribute to the building of the future. And this future is not the future of the cosmos but rather the future of my century, my country, my existence. . . . I belong irreducibly to my time."[40] It is here that we can pinpoint the major conceptual difference between prospecting and extractive mediation: extractive mediation is not just the construction of prospects but also the constraining of prospects. Extractive mediation is an alchemy that is projected through time. Decolonizing the deep sea requires that we give that time back, allowing a multiplicity of prospects for the many, rather than the few.

Putting mediation in conversation with extraction has the additional utility of revealing institutional and political complicities between activities that are typically envisioned separately as extractive or mediatory processes. We can see this, for instance, in the regulatory regimes for seafloor internet cables and pipelines. These two infrastructures, one explicitly envisioned as a media infrastructure and the other an extractive one, have historically been regulated in the same way. The Maltese and Swedish diplomat Avid Pardo (also known as the "Father of the Law of the Sea Conference") wrote in 1973:

> For some reason article 2 of the High Seas Convention does not distinguish between submarine cables and pipelines although the purpose of the former (apart from cables transmitting electricity) is international

communications and the purpose of the latter is essentially economic. . . . It is surprising that, at a time when increasingly numerous
pipelines are crisscrossing ever wider areas of ocean space, a distinction
between cables and pipelines has not been drawn and that no norms
have been proposed with regard to the construction, maintenance and
protection of the latter.[41]

Here, the imbrication of extraction and mediation can be described
in terms of the alignment between the development of two industries:
offshore petroleum and internet communications. My second and fifth
chapters push this point further, arguing that beyond this regulatory
slippage, both underwater media infrastructures and pipelines facilitate
simultaneous acts of extraction and mediation.

The political imbrications between corporate industries, sovereign
nations, scientists, and international regulators affects the public availability of information and data gathered through ocean mediations. Throughout my case studies, it becomes apparent that visibility is not necessarily
the goal of extractive mediation. In other writings I have developed a
concept of "dark mediation," which borrows from Eugene Thacker's language of "dark media" to describe how, contrary to mainstream efforts
to make the ocean more visible and accessible to humanity, military and
economic stakeholders have historically engaged in efforts to limit and
control signals in the deep sea in order to take advantage of this space.[42]
Unlike Thacker, however, I am less interested in the general condition
of inaccessibility or excess inherent in all mediation (what he describes
as "the mediation that cannot be mediated") than in how human agents
deliberately obfuscate information around particular environments like
the ocean. James Hamilton-Paterson writes that "among the sea's attributes are the capacity to conceal, the ability to stand for time, and the
quality of erasure."[43] Indeed, despite a prominent global movement to
make knowledge of the sea accessible, the deep ocean's remoteness and
physical inaccessibility has always been used strategically to shield con-
troversial or harmful activities in the ocean from public knowledge. Extractive mediation is therefore frequently a dark mediation that decouples
knowledge production and visualization from values like transparency or
visibility in its social manifestations. I discuss this most directly in chapter 3 with reference to the Deepwater Horizon oil spill, but to some

degree it is a condition that pervades all of the extractive mediations that I highlight in this book.

Returning to Jue's tactic of conceptually displacing media theory underwater, my approaches to dark mediation and extractive mediation also displace concepts around media, energy, and extraction underwater. There are important insights to be gained from studying extraction and mediation in situated terms, and specifically, from thinking about how these practices are enacted within our oceans. Mining is perhaps one of the most readily understood and imagined practices of extraction, yet because most mining activities happen on land, mining has become privileged as a metaphor or stand-in for other kinds of extraction, including, for instance, acts such as data mining or cryptocurrency mining. Offshore resource removals expand the extractive imagination to include not just acts of mining but also surveying, holding, diverting, filtering, and precipitating.[44] These, too, are culturally and technologically conditioned techniques of mediation that serve to convert ocean material into exploitable formats.

Tracing Networks of Meaning

Addressing the sociotechnological basis for extractive mediation in the deep sea requires a multi-pronged methodology—one that is capable of both pointing out extractive relations and moving beyond them. While I am interested in articulating histories, relations, and perceptions, the underlying motivation of this book is to support environmental justice in our oceans. I follow DeLoughrey's insistence that "it is vital to bring the theoretical discourse of the global north into dialogue with communities that both are at the forefront of present climate change and its historical survivors."[45] As such, although extractive mediation is a theoretical framework that can be applied broadly, I also aim to deliver specific, situated insights around the politics of ocean resource contestation that are more than theoretical, and happening on the ground today.

How do we build modes of reciprocity to oppose long-standing regimes of extractive mediation? Doing so requires attention to economic justice, epistemic justice, race and gender justice, and multispecies justice, as these often overlapping values and their corresponding injustices are interwoven in the design, planning, and implementation of ocean futures. Moreover, locating extractive mediation and its manifold impacts necessitates thinking about the complex processes of scientific translation across human and nonhuman worlds. Methodologically, I address this complicated set of concerns in two main ways: first, by interrogating the networks, institutional arrangements, and assumptions around deep-sea extraction, many of which thrive on their own obscurity; and second, by including the experiences of nonhuman agents, living and nonliving, in my case studies. My research methodologies are strongly influenced by an STS approach to social networks, which trace meaning and experience through a web of actors, as well as by historical methods such as media archaeology.

STS scholars such as Michel Callon, Bruno Latour, and Susan Leigh Star deploy critical methodologies that trace the networks and tools through which translation of information across social worlds is accomplished. For example, Callon's "sociology of translation" includes a recounting of the dissonance, power imbalances, frictions, and failures that accompany scientific knowledge. Although scientists are his guides through this narrative, Callon endows agency to nonhuman actors and emphasizes the instability and ephemerality of their associations. Scientific work is painted as a deeply political, social, and even linguistic endeavor.[46] Similarly, Susan Leigh Star and James Griesemer elaborated the frictions and cooperation inherent in doing scientific work, detailing the manner in which different actors in a group negotiate their heterogeneity. Star and Griesemer's mode of analysis is ecological, looking at the process of shaping and consolidating social links between humans and nonhumans in the constitution of scientific credibility. They build on Callon's notion of creating associations between actors, insisting on an antireductionist ecological model that allows for multiple viewpoints, multiple and intersecting social worlds, and simultaneous translations.[47] Each of these scholars allocates an important role for media: for Star and Griesemer they are "boundary objects" that synthesize meaning into coherent findings; for Callon they are devices of *intéressement* that perform the work of translation within a social network.

My discussion of ocean observations here is derived from interviews, site visits, archival research, and discourse analysis. I have talked with and learned from oceanographic researchers, attended oceanographic conferences, and visited labs and sediment archives across the country. This includes institutions such as the Woods Hole Oceanographic Institution, the Scripps Institution of Oceanography, the Oregon State University Marine Sediment Sampling Group, the Ocean Observatory Initiative's Cabled Array, and the University of Hawai'i Mānoa. Most of my field work was based in the United States, and as such I primarily critique U.S.-based rhetoric around global ocean observing. However, many of the extractive endeavors I examine are transnational and international, and perspectives on ocean regulation and observation from other countries are included as well.

By listening to researchers and examining video, photographs, maps, legal documents, oceanographic records, and other media objects, I not only gained an understanding of the technical specifics of how ocean expeditions operate but also zeroed in on debates and discussions internal to these fields at different historical junctures. This includes conversation topics ranging from underwater heritage to energy security, research funding, public transparency, and environmental impact. Reflecting the book's ethical commitments, however, I believe my critical embedding in the world of marine science has to be self-reflexive and reciprocal. As such, I acknowledge that as a researcher I am myself extracting stories and information from communities that study the seafloor. And yet here I hope also to build on those conversations with new insights, giving something back to the communities who have lent me their expertise.

My method of reading across scientific, industrial, and social practices and connecting each under the framework of critical media studies is a departure from previous work on deep-sea technologies, which has predominantly focused on individual fields such as nautical archaeology, marine biology, physical oceanography, ocean engineering, maritime literature, or art. While much attention is paid to scientific and engineering advancements in deep-sea exploration, fewer actors acknowledge the problematic role extractive industries play in determining what human relationships to the deep sea look like and will look like in the future. An STS approach to deepwater mediation allows us to see these webs of influence, incorporating the substantial political and ideological context for deep-sea media and technological development today. As I analyze the documentation practices of scientists, policymakers, and marine industries, I am also historically contextualizing mediation techniques and their assumptions and values. Extractive mediation is a discursive formation that is identifiable as a structure for knowledge in the oceanic archive. It reemerges today with a new relevance and with new frictions, manifesting in reversals to the rhetorical justifications for extraction from past eras. Thus there is a need to think about extractive mediation from a historical perspective as well as in terms of present and future prospects.

On this point, I am indebted to the technology-focused media archaeology of researchers such as Parikka and Starosielski, who consider the social, cultural, and material histories of media technologies. Michel Foucault describes archaeology as a method that plumbs the discontinuities and fragments of the archive, insofar as the archive exists as "the system that governs the appearance of statements as unique events. . . . [It is] the general system of the formation and transformation of statements."[48] Foucault later gravitates toward genealogical analysis, drawing on Nietzsche's theorization of genealogy as a mode that discovers but does not map.[49] For Foucault, genealogy complements archaeology by fixating on the historical contingency of different systems of knowledge, tackling not just the discontinuities between discursive formations but also questions of emergence, recursion, and reversal. (This is in contrast to histories that rely on tradition, continuity, or notions of beginnings and endings.)[50] So while archaeology focuses on structures and structural differences, genealogy is more process-oriented in thinking about the transitions between different discursive modes. Both methods are important to my media history of deep-ocean observing technologies.

Chapter Outline

Each of my chapters addresses a pairing of media technology and extraction, a piece of a pathway for the removal of information and material from the seafloor to the surface. I have selected case studies that capture these extractive mediations by maritime communities that have historically held and continue to hold enormous influence in the development of technologies used to monitor, manage, and ultimately exploit our oceans today. The structure of the book as a whole reflects an oceanic sense of temporality, exploring past futures and futures past. I begin with a vision of a world with vast resources from the past, in service of a future, and end by pondering a plan to physically remake the seabed in the image of our terrestrial transportation and communication infrastructures. Like roads crisscrossed by telephone lines, ocean infrastructures build on one another in both circular and sedimentary ways. Early notions of the seabed as a place of lost treasures inform the techniques and technologies that are eventually used to enable offshore drilling. This industry subsequently led a second proliferation of deep-sea media technologies, which made the extraction of seabed mineral deposits seem feasible. Seabed mining prospects in turn fuel new regulatory frameworks and further scientific research on benthic environments. The desire to foster and protect seabed investments today operates alongside an existing thirst to know the unknown, leading us to a full-blown colonization of and transformation of the seabed.

Chapter 1, "The Blue Archive and the Blue Frontier," begins this story at a place that many readers are familiar with: the recovery shipwrecks and buried treasure at the bottom. Following Jason Groves's reminder that we must pay attention to past inundations as we face future ones, I draw from the shipwrecks of antiquity to think about the potential wreckage of our future.[51] My entry point is an examination of how Chinese and American shipwreck salvage expeditions have been mobilized to make geopolitical claims and assert sovereign power, and I compare these developments to legal regulations of seabed mining. Deep-sea media innovations built for salvage and other archaeological needs contributed to a widespread imaginary of the deep as a repository of human technological pasts, largely excluding nonhumans from historical accounts. Yet to see the deep sea as a valuable and exploitable archive implied a sedimentary model of heritage that also embedded notions of human evolution, frontier conquest, and technological progress. This results in what I call a salvage-extraction dynamic, wherein notions of the blue archive and the blue frontier emerge in tandem.

This sedimentary model of history returns in chapter 2, "Swimmers in a Sonic Pipeline," which looks back at the history of petroleum seismology and the use of explosives to image the geological character of the ocean bottom. I interrogate the epistemology of the survey in particular, examining the ways in which sounding technologies, from early line surveys to air-gun arrays, adhered to an ideal of continuous and comprehensive information feeds. The desire for continuous sound images that could approach a photographic level of detail led to a bifurcation of the ocean's sounds into signal and noise, which, alongside economic motives to find offshore oil deposits, justified the emergence of a dangerously loud undersea environment. Thinking through sound studies as well as a multispecies perspective on sounding, I emphasize here that despite their presentation as innocuous forms of observation, these sonic counterpoints to drilling came at a steep ecological price and continue to imperil ecosystems and endanger species like the Rice's whale and the North Atlantic right whale.

Chapter 3, "Perilous Plumes," considers environmental impact through deep-sea mining. Unlike offshore drilling, the seabed mining industry is still in its infancy, and as such, studies of its environmental impacts are ongoing and are being attended to by industry and regulators alike. Much of the focus has been on studies of anticipated waste and sediment plumes in particular, which could potentially smother seabed life and carry toxic metals to distant areas of the ocean. Fixating on the figure of the plume, I discuss how attempts to model and predict plume behaviors have been central to environmentalist movements for decades, from studies of smokestacks to attempts to track oil spills. Taking into account the politics of visibility around buoyant fluids and aerosols, I consider how these emergent sediment plumes have mediated our imaginaries of environmental risk across temporal and spatial scales, feeding into a broader politics of extraction. I propose that the plume is helpful to think through as a metaphor for the seabed mining industry's development itself, which is marked by varying concentrations, turbulences, and the entrainment and detrainment of information and ideas between social realms.

The last two chapters look toward the future of ocean observing and international efforts to construct networked deep-sea infrastructure. Chapter 4, "Ocean Pacemakers," returns to the plight of whales, examining the expansion of satellite telemetry and the recruitment of marine megafauna as laborers within a planetary media infrastructure. While tags bring us closer to whales than ever before, these new glimpses into the in situ lives of marine organisms are not necessarily coupled to notions of empathy or kinship. Rather, I conceive of intimacy here as a condition of both precarity and possibility. Drawing on interviews with marine mammal researchers and tag developers on the U.S. Pacific coast, I demonstrate how cetacean telemetry has acted as form of multiscalar mediation that connects Western environmentalist ideologies to medical, commercial, and military cultures. As conservationist sentiments mingle with a mandate for data integration, both the whale and existing ocean-observation platforms are reoriented into a cybernetic framework for ocean health.

The final chapter, "Deepwater Feeds," continues the pivot from material resource extraction to data extraction, pointing to other modes by which depth is frequently viewed as a problem that can be solved through the technologization of nature. Starting with science fictional fantasies of *Homo aquaticus*, today's push to create a "wired abyss" through cabled seafloor observatories is often perceived as a major step in humanity's journey to become amphibious. My interviews and analysis of documents related to these emergent infrastructures lead me to pinpoint a tension between discourses of bioenrichment and biofouling: the first presumes interspecies interactions in the deep sea to be beneficial to humanity, while the second perceives aquatic species as obstacles or "fouling" agents when it comes to underwater technologies. The chapter's title draws on different valences of "feeding" to consider the transfer of both information and resources through these new seabed infrastructures.

Our media imaginaries of the deep ocean are often split along disciplinary lines. We associate scientific work with the look of discovery, popular media with spectacle or environmentalist concern, and industrial media with the look of speculation. The chapters in this book examine deep-sea looking and sensing relations without these divisions. From musealization, to prospecting, to modeling, I find that discovery, spectacle, and speculation often occur together. A healthy dose of technological determinism also acts as the binding agent between these modes of looking. But as this book will contend, ocean observing is not necessarily synonymous with high-tech methods, and the sustainability of these technologies is not a given. Extractive mediations are often walking contradictions. They imply both destruction and sustenance, life and death. In fact, to think about extractive mediation means venturing beyond binaries, beyond heroes and villains.

I think back to Jing Wei's story, where the ocean and our vengeful seabird are neither right nor wrong. At the end of the story, Jing Wei's struggle with the ocean is what creates the sandy beaches, a littoral space between land and sea. The dramas at our seabed similarly remind us that the oceans are connected to our lands. Ultimately, attending to these deepwater alchemies is not just about overcoming stormy waters to find stable surfaces. Instead, we find ourselves caught within the mixing and churning of the elements.

1

THE BLUE ARCHIVE AND THE BLUE FRONTIER
From Salvage to Extraction

> People who have been to the abyss do not brag of being
> chosen. They do not believe they are giving birth to any
> modern force. They live relation and clear the way for it, to
> the extent that the oblivion of the abyss comes to them and
> that, consequently, their memory intensifies.
>
> —Édouard Glissant, "The Open Boat," from
> *The Poetics of Relation*

The maritime museum is a haunted place. Displays of wreckage and
artifacts found at the depths do much more than guide us through still-
ness—they animate the past and articulate its relevance to the present,
breathing new life into old stories through annotation and audiovisual
mediation. In order to understand how the seabed became an extractive
frontier, I begin by looking into the past, at shipwrecks and buried treasures, for it was by first seeking what was lost that people began placing value on the ocean bottom as a space of profit. Steve Mentz coined
the term *Naufragocene* to describe the early modern period in which
shipwreck stories captured a certain crisis in Western understandings
of global culture: "Shipwreck resonantly names an epoch whose contours precede but also prestructure the arrival of fully global capitalist
exchange."[1] This sense of the Naufragocene as linking the seas and global
capitalist exchange continues today, and remains perceptible in the mediation of wrecks.

Mythologies of treacherous reefs and old wartime battle wounds,
reflected in sites like Pearl Harbor, the Mediterranean, the mid-Atlantic,

and the North Pacific tell stories of power and possession, the ebbs and flows of nations, and the ever-looming hardships and dangers of sea travel. The seabed contains traces of our old shipping routes, slave trade, and wars—a record of global conflict and imperialist expansion. There are an estimated three million shipwrecks on the ocean floor, from mesolithic dugout canoes, to Late Bronze Age vessels, to World War II submarines valued at sixty billion dollars.[2] Sharing space with these vessels are the decomposing, ossifying bodies of whales, mussel shells, gastropods, and man-made trash. Beyond the scenes of the maritime display is thus a question with a perpetual stake in the present and future: Who decides the value of this wreckage? How do we designate where (or when) ruins begin and end? What might we discover if we turn our attentions to the underlying logics, politics, and assumptions that order the dead, the dying, and the living?

When I first started visiting maritime museums, it was as an afterthought—a way to spend the time in between my archival research and writing on oceanographic technology. But while this started as a passing curiosity, I soon came to see shipwrecks as consequential players in the maritime world as a whole, a key piece of the story of deepwater exploration. Touring old frigates, reading displays, examining rusted objects, and watching live-streamed nautical archaeology videos, it became clear to me that our mediations of the seafloor as an archive of cultural heritage are far from innocent.

Two examples stood out to me—one from China and the other from the United States—both of which seemed to reflect on the political tensions between the two powers. The first is the Guangdong Maritime Silk Road Museum, a site that recalls China's global commerce in a way that directly justifies China's present-day endeavors to control the contested waters of the South China Sea. The second is the Pearl Harbor National Memorial, which returns to a moment of trauma to project an ideal of the nation, frozen in time. A threshold between wilderness and civilization, the debris-strewn repository of the seafloor is routinely mined for potsherds of the past in the service of ideals about civilization, progress, power, and statebuilding.

In this chapter, I home in on the mediated imaginaries of the seabed as an archive of technological feats and failures, interrogating the entangled temporalities of salvage and extraction that underpin the way in which the seafloor is regulated, imagined, and ultimately transformed by anthropogenic operations. My aim is to consider the ways in which notions of heritage, of commons, and of conquest intersect with the modern-day exploration and extraction of seafloor resources. I ask, how has an understanding of the ocean bottom as a cultural archive helped to produce the seabed itself as valuable and extractable? How do evolving media techniques for archaeological excavation influence the ways in which this blue archive is given historical meaning? And how might heritage and inheritance be rethought, in the context of a living (and dying) ocean?

To answer these questions, I will chart a path from nautical archaeology, to frontier discourse about the seafloor, and finally to my own concept of a living archive, which rethinks the role of inheritance and heritage in dictating human relations to the ocean. Specifically, my argument begins with a consideration of how cultural heritage is constructed through the digital and analog mediations of drowned relics to secure a linear perspective of history. I then explore the ways in which narratives of shared heritage that accompany archaeological efforts act in concert with the "recovery" or extraction of seafloor resources, with particular reference to the legal regulation of the deep seabed. Bringing notions of heritage and frontier imaginaries of the deep seabed together, I propose a *salvage-extraction dynamic*, in which ideas of the archive are superimposed onto the seabed's extractive possibilities, producing it as a static reserve through which civilizing narratives about nature, culture, and heritage emerge. The epistemological collapse between material and informational extraction is an instance of what I refer to here and elsewhere in the project as extractive mediation. I will end by considering the liveliness and changeability of this blue archive, exploring alternative approaches to curating futures from the seabed.

What I aim to do here is to show that far from being an academic pursuit detached from capitalist endeavors, archaeological salvage at sea has, in fact, much in common with extraction. On the one hand, nautical archaeology subscribes to frontier-oriented notions of progress and control over wilderness, while on the other, extractive industries like seabed mining frame themselves through common heritage. Using the Guangdong Maritime Silk Road Museum and the in situ memorialization of the Pearl Harbor wrecks as case studies, I will make the case that the production of archives from the seafloor is itself a contested activity that produces meaning not only in terms of our inherited past but also as an outcome of human capitalist excess. The result of this is that the blue archive itself is transformed into a resource to be mined. Put simply, the blue archive and the blue frontier are two sides of the same coin.

This unique understanding of the deep sea as both archive and frontier, as a reservoir for preservation and progress, is achieved not by one actor, but by many. Archaeologists, politicians, scientists, industrial contractors, and lawmakers each play a role in configuring the seafloor both materially and semiotically, delimiting its value to human civilizations. Ultimately, my proposal of a salvage-extraction dynamic means interrogating the rhetoric at the heart of emergent underwater practices to consider how various actors work together to produce lasting structures for knowledge and meaning. I take an interdisciplinary approach to these mediated imaginaries of history, reading texts from nautical archaeology, international policy, and oceanography in order to understand the cultural reach of various blue archives. I will examine how this network of actors defines convincing evidence of heritage, how they interpret relics, and ultimately, how a "negotiated order" between scientific practices, cultural and social context, and political agendas comes to mediate this seafloor museum.[3]

The Blue Archive

Writing on Pearl Harbor, James Hamilton-Paterson puts forth that shipwrecks traverse four categories: "tombs, time capsules, gold mines and time bombs."[4] I pondered these categories of the wreck as tomb, archive, time bomb, and gold mine in 2019 during my own visit to Pearl Harbor, where the ocean is mediated as a memorial and a repository of knowledge about U.S. history. Pearl Harbor contains traditional brick and mortar displays of the past, but it also makes use of war graves—in situ memorializations of wreckage from the USS *Arizona* and USS *Utah*. To reach the USS *Arizona* memorial, visitors must board a shuttle boat, which takes groups to a structure built directly above the shipwreck. Inside, there is a gaping hole where visitors can peer into the murky green depths of the wreckage, and spot colorful fish flashing by. These curated experiences are designed to evoke both a sense of remembrance and the novelty of discovering a buried and valuable past.

In watery graves such as these, Hamilton-Paterson's assertion that the seabed is a space of oblivion frozen in the past and that "whatever it engulfs becomes ancient almost immediately" exists in frictive tension with the ocean's nonhuman temporalities, as well as its recurring positionality to the human world as a technological frontier and a site in which the global modern is imagined.[5] The nonhuman agencies at work are part of the Pearl Harbor archive and arguably more perceptible to the average viewer than the wreckage itself. Meanwhile, off to one side of the *Arizona*, black drops of oil periodically squeeze out of the hull and swirl into the surrounding water. This old wound was kept for its emotional poignancy—an ongoing oil spill representing the still-dying wreck. More than an ordinary tomb, the memorial simultaneously presents us with an ongoing death and a growing community of life. The juxtaposition of the two temporalities seems to complicate Hamilton-Paterson's suggestion of discontinuity or stasis, in the way that it simultaneously reveals the passage of time.

These details, alongside costumed tour guides and other, more explicit attempts to put visitors in the shoes of old crewmen, rely on animation as a form of remembrance. The dead are once again made lively for visitors, speaking through placards, audio recordings, tour guides, and visual displays. Sara Rich speaks about a "shipwreck hauntography," arguing that nautical archaeological endeavors often imply an imperialist savior paradigm in which wrecks are "resurrected."[6] Indeed, some artifacts, like *Nanhai No. 1* wreckage at the Guangdong Maritime Silk Road Museum, are almost explicitly resurrected in the service of modern-day geopolitical conversations.

Peter Campbell points to a systematic way in which marine artifacts are exploited as political tools to broaden territories. Russia, for example, has led several high-profile nautical expeditions, with Vladimir Putin himself participating in archaeological dives in the Black Sea to explore ancient shipwrecks near Crimea—an object of a territorial dispute between Russia and Ukraine—prior to Russia's 2022 invasion of Ukraine.[7] Putin mentioned the purpose of the expedition was to "understand the development of ancient Rus's relations with its neighbors, as well as the development of Russian statehood."[8] Such media stunts are transparently about producing a politically advantageous narrative about nationhood. Through state-sponsored expeditions in nautical archaeology, countries routinely translate the vast artifacts of exchange scattered across the seafloor over centuries into cohesive pictures of ancient trade routes—a performative endeavor that has, in many cases, clear geopolitical benefits in the present day.

These examples speak to a much longer disciplinary history in which archaeology has served as a political tool both in and out of the water. For instance, in her examination of archaeology and the Israeli state, Nadia Abu El-Haj discusses the role of archaeology in reinforcing Zionist settler nationhood. Couched in a research agenda and epistemology that assumes a specific idea of nations, ethnicities, and historical emplotment, El-Haj contends that this "critical tradition [of archaeology] is united, at its most basic level, by a commitment to understanding archaeology as necessarily political."[9] While the underwater context differs from the terrestrial archaeology in its question of settler colonialism (people do not literally live underwater), aquatic dominance nevertheless equates to an extension of national identity, political influence, and economic power over spaces that provide the livelihoods of many human and nonhuman communities. Similarly, in his examination of Mediterranean maritime museums, anthropologist Naor Ben-Yehoyada argues that struggles over certain recovered underwater relics such as the *Dancing Satyr* point to a process of "kinshipping," a "reaching back to the distant past to inform and then change present political relations."[10] Here, Mediterranean transnational identity is reinforced through a reconciliation and continuity between ancient maritime connectivity and modern-day forms of connection.

Cases such as these support the idea that shipwrecks continue to capture a certain crisis of cultural authority, as well as the authorities with which culture subtends (political, legal, industrial, scientific), well into the twenty-first century, just as they continue to structure global exchanges of capital and understandings of global futures. In an era of climate change—and, by proxy, of the ever more precarious geopolitical contestations for natural resources—the moment of oceanic awakening named by the Naufragocene continues onward as we venture into deep waters. Historian Joan Scott explains that historical subjects write themselves into histories in order to retrospectively stabilize identity.[11]

This is precisely what happens when archaeologists endeavor to reclaim what is lost; archaeological narratives include traces of the present day, either in their redemptive undertones or in their political conveniences. Narrative operations around shipwrecks are a kind of fantasy in which collective identity is secured through the resolution of antagonisms and gaps.[12]

This dynamic has been explored by historians in relation to the infamous example of the 1912 sinking of the *Titanic*. The sinking of the *Titanic* captured an anxiety about a future displaced to the past, which then animated musealization efforts, in what Andrea Huyssen describes as an impulse to construct "present pasts."[13] In other words, as a high-tech failure, the wreck itself thwarted an assumption of linear temporality which was then rewired and resolved through musealization. Greg Siegel echoes this, describing the anxiety in slightly different terms: "An accident too horrible to ignore, too devastating to discount, the *Titanic* seemed to offer startling proof of *progressus interruptus*, of forward movement 'flinched.'"[14] In this version of the story, the *Titanic* comes to serves as a warning for future generations, "a cautionary tale about the perils of human hubris."[15] The event marked a cultural moment in which there was widespread fear and ambivalence toward ocean depths as a whole and of its capacity to thwart ocean travelers. As one 1932 article from the Submarine Signal Company puts it: "The bane of the mariner is really the bottom of the ocean. How to keep off it is his ever present problem. . . . Back in the mind of the shipmaster and sailor is the haunting fear that the craft they navigate will reach the bottom either by sinking, through collision or storm, or by grounding in shoal water."[16] Viewed in this light, the oceanographers who helped retrieve the *Titanic* were forward-thinking heroes willing to swim toward the watery abyss of past failures, a space that struck fear in the hearts of others.

Chinese shipwrecks are an apt case study for how wrecks can be used to resolve antagonisms and gaps, as the state has been using nautical archaeology expeditions to bolster claims of global dominance and ownership over ocean space through the shipwrecks that shadow trade routes. At the time of writing, brewing tensions between the United States and former Eastern Bloc have boiled over in a number of ways, from accusations over the mishandling of the Covid-19 crisis, to protests over neocolonial policies, to the more recent Russian invasion of

Ukraine in 2022. But while these conflicts over land, resources, and power may seem new and explosive when encountered over breaking news or viral social media discourse, we can also see their long-brewing traces in the less spectacular maritime realm.

Perhaps the most overt example of historical subjects writing themselves into history is the Chinese development of the idea of the "Maritime Silk Road," articulated most elaborately by a museum in Guangdong that opened in 2009 and showcases underwater archaeological relics related to China's foreign trade. The location of the museum overlooks the Pearl River, and is next to the waters deemed a central part of the Maritime Silk Road. The very first of these relics recovered by Chinese underwater archaeologists was the *Nanhai No. 1* shipwreck, a merchant ship transporting porcelains as far back as the Song dynasty (1127–1279). It was found in 2011 at the mouth of the Pearl River, which was then deemed the starting point of the Maritime Silk Road.[17] The ship's name, *Nanhai*, translates as "South China Sea." Throughout news media released about the excavation, there is a clear nationalist call to extrapolate the significance of the shipwreck, with some even comparing it in significance to the Xian terracotta warriors.[18] For instance, archaeologist Xu Yongjie maintains the Chinese perspective that *Nanhai No. 1* should be seen as synecdoche, a part of a whole: "窥一斑而知全豹," a "Chinese idiom [that] means peering at one spot and knowing everything."[19] Xu ends with a reaffirmation of the trade route as a part of China's proud past: "We can confidently predict that as the excavation proceeds, the shipwreck will provide much more new evidence to help us appreciate the past prosperity of the maritime silk road."[20]

In 2015, the Chinese State Administration of Cultural Heritage launched another massive archaeological expedition to recover artifacts in the South China Sea. The press release reads, "Our ancestors have been producing and living in the South China Sea since ancient times. The Xisha (Paracel) Islands are an indispensable part of the 'Silk Road on the Sea' route, leaving behind a great amount of underwater cultural heritage from various dynasties."[21] China's recent interest in nautical archaeology has focused on the discovery of relics along areas like the Xisha archipelago, part of the contested South China Sea territory, which holds oil and gas natural resources and fisheries and is a key strategic area for military operations. Pushing back against territorial claims by Taiwan and Vietnam, Chinese press make heavy-handed assertions connecting heritage artifacts to territorial ownership, thereby claiming sovereignty over an extended space.[22]

It is hard not to see the present-day benefits of the historical narrative created by the museum; in 2013, China announced its modern-day Maritime Silk Road Initiative (MSRI) alongside the Silk Road Economic Belt (SREB) as part of its "One Belt, One Road" project. The initiative focuses on infrastructural connectivity (highways, railways, ports, power grids, pipelines, and telecommunications networks) between China, Southeast Asia, India, Africa, and even as far as the South Pacific.[23] Whether seen as a step toward Chinese global hegemony or a step toward economic integration, the geopolitical stakes of MSRI remain high and are no doubt bolstered by the historical fantasy manifested in Chinese nautical archaeology endeavors.

Mediating the Blue Archive

There are two theme areas to the Maritime Silk Road Museum: an exhibit area for the *Nanhai No. 1*, and an area for other artifacts of the Maritime Silk Road. The layout is composed of exhibition halls in interlinking elliptical rings, the largest palace being reserved for the shipwreck (see Figure 2). It includes a twelve-meter-deep aquarium containing the steel-lined well, which aims to replicate the benthic environment where the ship sank. Other exhibits show wreck parts and daily sailor's articles and trade items. Display boards, videos, VR experiences, and other forms of multimedia adorn the exhibition halls.[24] But the real draw of the museum is not necessarily the rooms full of porcelains so much as the live archaeological show, where museumgoers can watch workers sifting through the artifacts. Macao Museum of Art director Loi Chi Pang emphasized the scale of the intact salvage process itself in an interview about *Nanhai No. 1*:

> It was not an easy exhibition to arrange. The significance was in the recovery and archaeology, a first in China. Our mission was to explain that to the public, how this had been achieved. . . . The intact salvage of Nanhai No 1 was an enormous feat of engineering and proved extremely difficult. It involved fighting against the elements and the mud surrounding the boat thick with the sediments coming back at

the speed of 10cm a month, as well as the instability of the weather. The operation was followed closely at home and abroad.[25]

Xu explains that the excavation included surveying and mapping the ship and creating a virtual test pit from the data, prior to the test pit work itself. Muds were dusted off and analyzed, artifacts were numbered and recorded, and the whole process was put under the camera from a birds-eye view. Ultimately, in showcasing the logistical spectacle of archaeology, the tone of the museum is not one of mourning; rather, the *Nanhai No. 1* excavation is cast as a national achievement. The act of archiving wrecks like the *Nanhai No. 1* becomes a media endeavor insofar as it entails communicating and translating information about human and natural histories. Whether it involves filming, preservation, or musealization, archival mediation acts as a template for important ideological, political, and industrial commitments.

Media archives have typically been theorized as the stuff of paper, decaying celluloid, and more recently bits and bytes.[26] The blue archive as I discuss it, however, comprises what is in reality a concatenation of

Figure 2. The wreck of the *Nanhai No. 1* is housed in a water tank within the Guangdong Maritime Silk Road Museum. Audiences can observe its excavation through a glass wall. Photograph by Zhangzhugang, 2013.

several different kinds of archives, beginning with an understanding of the seafloor itself as a natural archive of the past—an elemental medium, as scholars such as Jay Bolter, Richard Grusin, and Alenda Chang have articulated.[27] Archiving consists of porting information from one context to another; archives are simultaneously an abstraction and materialization of previous media. One way to think through the object fetishism of maritime museums is therefore to consider their mediated displays of objects in relation to other formats and venues for aquatic archiving, such as a marine geological archive. If the shipwreck museum is one cultural space in which the ocean bottom is made into an archive, marine sedimentary records are another type of "blue archive" buttressed by closer ties to scientific authority.

Éduoard Glissant once proposed sedimentation as a metaphor for historical processes at large,[28] and indeed, from a scientific perspective, the prepositional equation of a sedimentary "bottom" to temporal beginnings appears intuitive. Scientific fields like paleontology, geology, and archaeology establish that fossil records and stratigraphic data preserve and remediate information about Earth's history as well as the history of life itself. I paid a visit to Professor Joseph Stoner, a paleomagnetist and geologist at Oregon State University and director of the OSU Marine and Geology Repository (OSU-MGR), to learn more about how this seafloor geological archive is produced, managed, and interpreted. Strolling through the refrigerated core repository on the OSU campus, we gazed upon stacks of long tubes of sediment, each labeled with dates, locations, and other metadata. Stoner explained to me that geological cores retrieved from the seafloor can be read like "time machines" or "a solutions manual to the Earth."[29] His perspective clearly frames these geological objects as media objects, implicit in the metaphor he offered up to me:

> What you need are records that preserve the geomagnetic field really well, that accumulate at high rates so you don't smooth out too much of the information, so you can really see a clear high fidelity picture. It's like going from an old snowy TV screen to a 5K monitor where the clarity of picture just becomes greater.[30]

Rocks become pictures, and sedimentation rates equate to fidelity. The marine core repository is thus a media format that seeks to make legible sedimentation itself as a historical process. We dig *down* to move *backwards.* To get to the bottom of our pasts, we get to the bottom of the sea.

To describe this sedimentary view of history and its connections to extraction, I find critical purchase in examining anthropologist Elizabeth Povinelli's revision of the concept of biopower in "geontopower," or the tactics in late liberalism to maintain distinctions between life and non-life (*geos*): "Geontology is intended to highlight, on the one hand, the biontological enclosure of existence. And, on the other hand, it is intended to highlight the difficulty of finding a critical language to account for the moment in which a form of power long self-evident in certain regimes of settler late liberalism is becoming visible globally."[31] Povinelli's geontopower encapsulates fields like geology and archaeology that, in many respects, see themselves as dissecting death, thereby re-creating a boundary between the nonliving past and the living present through the spatial boundary of surface/seafloor.

Like geologists, nautical archaeologists have long understood the seabed to be a palimpsest of human cultural pasts, giving the geological subject a historical corollary. The underwater context thus mirrors the terrestrial in its perpetual obligation to save a wreck or to "get to the bottom" of its demise. Archaeology justifies the enactment of vertical power, from the surface to the seafloor's lowest layers. Yet this very tendency to think in terms of historical sedimentation creates space for extraction, which likewise depends on the distinction between life at the surface and a static, nonliving floor. We might alternatively call this "geopower," as Elizabeth Grosz does, to describe a "capitalization of the forces of the universe" that intersects with but is not reducible to political potentials.[32] The sedimentary perspective on history and the value afforded to sedimentary material itself thus points to a fluid, dialogic relationship between the geopolitics of excavation and a geontopower of extractive capitalism at large.

Consider now that the official Maritime Silk Road Museum website states, "The success is an unprecedented achievement and really a landmark in the history of world underwater archaeology. The wreck which has slept in the seabed for over 800 years, now revived."[33] To see the seabed as a bed where sleeping ships lie means to think of it as a frozen space, awaiting human intervention to be awakened or enlivened. This blue archive as constituted by the Maritime Silk Road Museum is therefore one that does not quite replicate but does resemble older, more fearful understandings of the ocean bottom as a graveyard. However, unlike Povinelli's version of geontopower, which reproduces a life/death binary, Chinese nautical archaeologists are not just dissecting death at the bottom; they are re-creating life, asserting the power to create continuity between past, present, and future, with of course, a particular vision of civilization in mind. Rich's "shipwreck hauntography" helpfully frames these top-down perspectives of salvage as a site of uncanny resurrection. The same could be said for mining and drilling efforts. Indigenous ontologies of fossil fuels often position them as former kin, dead plants and animals long gone.[34] What is mineral extraction, if not a hunting and trapping of ghosts?

Archaeology produces its material culture through the technical work of excavation—a culture that is "embedded in the terrain itself, facts on the ground that instantiate particular histories and historicities."[35] The Institute of Nautical Archaeology asserts that expeditions are conducted to "increase knowledge of the evolution of civilization through the location and excavation of underwater sites."[36] This invocation of "evolution" is suggestive. It points to a fundamental worldview of human–nature relations in which civilization's evolution is read through progressive (as opposed to ongoing or mutually constitutive) human attempts to traverse and control natural environments, regardless of whether or not those attempts are successful. The deepest layer, then, is also the most "primitive," destined to be usurped by the more evolved technological remnants of man. Meanwhile, technological configurations capable of reconstituting the lost wreckage of the past are seen as highly evolved.[37] The archaeological diagnosis, dissection, and analysis of deep-sea ruins thereby contributes to a growing pool of recorded historical knowledge that reinvigorates a story told by modern society about its evolution.

Media theorists Wolfgang Ernst and Jacques Derrida have both suggested that the capacity to delimit what is archivable as well as what is forgettable is a constituent part of what defines the archive.[38] In other words, to see the seabed as an archive of such stories first is to have in place a structure in which objects may be defined as archivable.[39] A Derridean perspective would see the seabed itself as archivable to the extent that fields like nautical archaeology apply systems of management to it. For the seafloor, we define not only what counts as a valuable piece of heritage but also how this piece of heritage should be preserved. It is a filtering on two levels—both in terms of narrative and a material cleansing of the object itself. Ultimately, the ability to determine the contents of a such a cultural archive equates to the power to claim indexical or evidentiary truths about the past. This is how creating the archive becomes an act of mediation that anticipates its applications for the drawing and extension of national borders.

Media tools and standards not only produce the sedimentary blue archive as such, but also play a crucial social role in delimiting the credibility and competency of the professional field and establishing its privilege over alternative knowledge claims or cultural perspectives.[40] Satellite imaging, now a popular component of the archaeological gaze, is perhaps the quintessential example of media technology facilitating geopower over environmental depths. Writing about the excavation of Cleopatra's palace, Lisa Parks argues that as archaeological tools, "satellite images frame the earth as a massive excavation site waiting to be plumbed. . . . In treating the earth's surface as a script, archaeology imagines the planet as the raw material of the ancient past."[41] But satellites do more than just read surfaces. Increasingly, satellite sensors detect beyond the visual range with "microwave, infrared, and radar-imaging sensors" that "can pierce clouds, jungle canopies, sand, and even soils."[42] For Parks's case study, satellites did not merely improve archaeological vision; they penetrated nature as a feminized unknown, enabling Western cultural discourses around Cleopatra that played up the unveiling of sexual spectacle and racial ambiguity.[43] By asserting human agency over sedimentary pasts, archaeologists produce historical narratives that, in this case, strengthen hegemonic perspectives not only on ancient civilizations but also on technology as a locus of agential power.

Contemporary notions of environmental mastery depend on tools of mediation like these satellites and cameras, as well as on more hands-on methods of excavation. At sites of excavation, archaeologists, like other ocean scientists, define features and objects of relevance and then work to preserve them. Of course, some elements must be excluded—especially those that challenge the desired linear notions of progress or development. "Seawater asks us to rethink terrestrial notions of the archive or database as informed by the language of earth and sediment," writes Melody Jue, "and instead consider them in terms of seawater's capacity for protean transformation."[44] What do we do, for instance, about something like the 1881 Kingston shipwreck, which is now home to forty-eight species of corals?[45] In the case of shipwrecks, we encounter the paradox of the ship of Theseus made literal: If Theseus's ship is decaying and every plank is replaced with a newer and stronger timber, is it still the same ship?[46] Ultimately, what is meant by words like "preservation," which describe the transformation of seafloor debris into members of a cultural archive, hinges on multiple guiding principles for both restoring the ship and minimizing future wear and tear.[47]

Material preservation of most shipwreck artifacts is focused on both the prevention of future degradation and extensive restoration to return the artifact as close to its original (pre-wreck) state as possible. This involves scrubbing all traces of nature, or the action of natural history on the shipwreck. Specifically, a major part of shipwreck restoration involves the removal of salts, stains, and other mineral deposits, followed by a drying process.[48] A partial restoration process, however, can also allow for other omissions. The careful cleaning and reassembling of ship anchors such as that of the SS *Clan Ranald* have at times served as metonyms for the ship itself, figuratively anchoring the wrecks in a fixed temporal place and comfortably leaving the material event of sinking behind.[49] In each case, the production of an archived shipwreck separates it from the fluidity and turbulence of the ocean. Salvage thus typically enacts nature/culture divides, casting seawater as a breed of vulture. From this perspective, salvage is also almost always a "race against time."[50] The cannibalization of artifacts by marine life is the index that allows us to see the passage of time, while extraction becomes an intrusion that resists this natural time.

The deep seabed as traditionally narrativized by archaeologists is thus a "ground zero" of history: a space that intertwines nature and civilization, past and future. It adheres to a teleological model of progress and expansion that is tied to a need to redeem the past for the sake of the present. With salvage-extraction, to retrieve from the archive is to remember, insofar as re-membering is a stitching of a lost moment in time back into a controlled time-space of human history. Above all, shipwrecks are understood to *already belong*, whether to a nation, a company, or to humanity as a whole. Everything that is excess—rust, waste, enterprising corals, and the cyclical upheavals of nature—is more readily discarded. As such, the blue archive comes to appear static because of its adherence to linear human histories, and its erasure of natural ones.

And yet, even with these material cleansings, there is a foreboding twinge—a recognition that more shipwrecks loom on the horizon. As the seas rise with climate change, this foreboding fear takes on a different tone. As Elizabeth DeLoughrey puts it, "the ocean as medium can symbolize the simultaneity or even collapse of linear time, reflecting lost lives of the past and memorializing—as an act of anticipatory mourning—the multispecies lives of the future of the Anthropocene."[51] This specter of recurring, future wreckage gives us the pivot from thinking in terms of heritage to speculating about risky frontiers. As I demonstrate next, archival practices and frontier-oriented exploration are not mutually exclusive. There are fluencies between the archaeological impulse and mineral extraction, as two related salvage-extraction activities. This is evident again in legal language; in wreck videos that both memorialize lost lives and provide anticipatory glances at a still-wild yet resource-rich blue frontier, and in industrial resource extraction itself.

From Common Heritage to Blue Frontier

Human beings leave behind a great many traces in the ocean, but only some are deemed "cultural heritage" with "nonrenewable" value and thus worthy of "recording, preservation, and responsible management."[52] This is why, in legal terms, the designation of objects as cultural heritage has become a flash point for the research community seeking to mitigate competition from opportunistic treasure hunters. Reflecting the capitalist resource conversion at the heart of extraction, the 2001 UNESCO Convention's definition of underwater cultural heritage emphasizes the relationship between heritage and property, stressing "the necessity for cultural heritage to be owned and regulated in order to be safeguarded."[53] This intersection between ownership and safeguarding is easily deployed in territorial regions to advocate for the protection of archaeological artifacts. However, in nonterritorial regions like the deep sea (known in legal parlance as "The Area"), the third UN Convention on the Law of the Sea (UNCLOS) stipulates something slightly different. Under this law, deep seabed archaeological sites can be interpreted to fall under the "'cultural' *common* heritage of mankind so as to include sites found on the seabed beyond national jurisdiction."[54] The seabed thus assumes the inheritance of something by humanity as a whole, language that is usually reserved for regulating the exploitation of natural resources in this region. The term "common heritage of mankind" (CHM) is typically used to argue for the protection of mineral deposits "inherited" by humanity—a point that I shall discuss later on.

The explicit language of the convention states: "All objects of an archaeological and historical nature found in the Area shall be preserved or disposed of for the *benefit of mankind as a whole*, with particular regard being paid to the preferential rights of the State or country of origin, or the State of cultural origin, or the State of historical and archaeological origin."[55] As Anastasia Strati argues, however, this stipulation is vague and prone to disputes. The parameters for what demarcates objects as worthy of protection or for how to determine the state of historical and archaeological origin when geopolitical territories shift over time remains unclear.[56] Ultimately, to make a case for historical preservation, nautical archaeologists make claims about heritage that take into account present-day political relationships and future profits. These debates speak to the geontopower of nautical archaeology. As my opening examples attest, the salvaging of shipwrecks has always been a political endeavor, extending the historical relationship between oceanic mastery and colonization from the fifteenth century to the present day. These values further enshrine a practical consideration of the seafloor as a frontier space.

The seafloor is globally mediated as both a resource frontier and a technological frontier. Both senses of the blue frontier link up to ideas about common heritage. The classical orientation of frontierism, as Jody Berland describes it, includes narratives of progress and evolution.[57] These ideas build from the frontier of the bygone seventeenth- to nineteenth-century American West, which was largely responsible for contributing notions of futurity built on colonial dominance, violence, and exploration. It was Frederick Jackson Turner who put forth the thesis that the shifting line of the frontier represents "the outer edge of the wave—the meeting point between savagery and civilization."[58] For Turner, the harsh environment of the American frontier shaped national identity, uniting citizens through the conflict and struggle against wilderness. The domination of nature and primitive men by frontier men thus acted, in Turner's estimation, as the engine of progress.

The imaginary of the deep seafloor as a lawless frontier of a similar kind has been expressed through writings over hundreds of years, from nineteenth-century authors like Jules Verne to present-day territorial contests. Historically, the process of civilizing frontiers has also often manifested in resource extraction, mining, and drilling, enacting colonial power and capitalism through processes like neoliberal privatization, or the casting of natives as obstructions.[59] Indeed, we see this story play out again and again in remote natural spaces. Less than one-thousandth of the deep ocean has been studied by scientists,[60] yet its landscape, constellated by hydrothermal vents, seeps, mineral formations, and biodiverse communities of clams, tubeworms, crabs, bacteria, and other organisms, has galvanized exploration by industrial, scientific, and political actors alike. Indeed, the old epithets of frontierism come through strongly in articles about seabed mining, which describe the deep ocean with terms like "silent worlds," "alien ocean," "final frontier," "invisible frontier," and "new frontier."[61]

With regard to the seabed area beyond the jurisdictions of individual nations and exclusive economic zones, the International Seabed Authority (ISA), created through UNCLOS, is the regulatory body responsible for granting seabed mining licenses to 159 countries. This space is governed under the principle of CHM, where it becomes simultaneously a global archive and a resource common through the universal concept of heritage for all humanity. More specifically, according to UNCLOS, the ISA acts on behalf of "mankind as a whole" by ensuring that financial resources get split equitably and with the interests of developing states in mind, shadowing the UN's language on archaeological regulation.[62] This legislation largely responds to the fears imagined through Hardin's tragedy of the commons, or the idea that individuals, without regulation, would despoil a common landscape or common resource out of self-interest and capital accumulation. In the blue archive, heritage is thus a concept that accords value to that which is sedimented, redefining the seafloor as extractable.

Ideas about preservation, disposal, and management of circulation coexist in legal language. The commons is not pitted against enclosure, but rather gets integrated into the management of capital flows. More than merely standing in for a relationship to the past, this notion of common heritage takes into account present-day political relationships and future profits. It comes as no surprise that both the ISA and the United States' Deep Seabed Hard Mineral Resources Act, administered by NOAA, describe seabed mining in language resembling that of archaeology. "Exploitation" is defined explicitly by the ISA as "The recovery for commercial purposes of mineral deposits in the Area and the extraction of minerals there from."[63] Like shipwrecks, such resources are recovered because they *already* belong. And, in a related vein, the ISA regulations on prospecting for ferromanganese crusts ends its list of regulations with a mandate for prospectors to report "any finding in the Area of an object of actual or potential archaeological or historical nature and its location," refitting every mining expedition as simultaneously an archaeological one.[64]

Crucially, however, while CHM espouses inclusivity and democratic ideals, it delimits heritage in a way that opens the seafloor up to contested claims on land and conflict with indigenous communities. A salient example of this imperialist co-optation of CHM was the passing of New Zealand's Foreshore and Seabed Act of 2004, which declared the Crown the owner of the country's foreshore and seabed. While the act was passed in the name of "common heritage of all New Zealanders," practically speaking, it bulldozed over traditional Māori property rights and naturalized state appropriation of their lands. The act was effectively a "'sea grab' by the state that disenfranchised Māori from their customary title."[65] As DeLoughrey observes, the construction of the seabed as a commons in this instance allowed settler colonies to erase the indigenous subject and make a claim for legitimacy. She argues that resisting this narrative would involve challenging the "(geontological) ground on which the state derives its sovereignty, including the state's claims to the strand seabed, and creatures of the ocean as a 'common heritage' and thus political territory."[66] Perhaps, Columbus O'Donnell Iselin, one of the few oceanographers writing rather pessimistically about the possibility of seabed mining, put it best in 1968: "Future uses of the deep ocean are far from being bright. It will not be easy to put them to use for the benefit of all mankind."[67] The construction of a speculative seabed archive through the language of common heritage can thus, practically speaking, become a tool of colonization. In the blue archive, the notion of a "resource" or "cultural artifact" is thereby invented alongside the designation of others as obstacles (ocean waste, natural turbulence, indigenous communities, environmental fragility).

The salvage-extraction dynamic inherent in CHM shifts the focus of heritage from the mere preservation of objects in space and time to their distribution. "Extraction" becomes "recovery" at the same time that recovering objects of heritage in the seabed comes to presume a strict regime for extraction, management, and circulation. Although the traditional prerogative of cultural heritage is to preserve time and save historical artifacts from oblivion, the prerogative of common heritage is to manage extraction while assuming that many actors are already competing to salvage (or recover) objects of value. The seabed archive thus becomes not simply a holding place but also a resource to be mined. If minerals can be reframed as part of human heritage, then heritage itself can be commoditized. Monetary futures are enabled by the slow buildup of gold and copper nodules over thousands of years and by the slow recovery of valuable wreckage over centuries. Extractive temporalities of capital and temporalities of salvage thus merge.

With its universalizing of human experience and its economic underpinnings, CHM ultimately reproduces a paternalistic view of nature, focusing on the preservation and recovery of objects of value, while the effects of salvage or extraction in a fluid and mobile environment become secondary obstacles. Frontier narratives like that of mining tend to relegate the nonhuman or the indigenous to that which must be dominated or controlled. To men like Turner, the "outer edge of the wave" is a violent, unruly site of struggle, where wilderness is civilized. But acknowledging the role of heritage in constituting frontierism in the deep requires a rethinking of the wave itself, the zone where nature and man supposedly make each other.

Richard White speaks of two dueling imaginaries pertaining to the old American frontier: one is defined by the figure of the scout, as popularized by Buffalo Bill Cody, and the other by the figure of the farmer, shepherded by Turner. The scout is defined by conflict with native peoples, while the farmer seeks to tame the natural world; both perform a version of flag planting through acts of domination and control. White's discussion of these dominating narratives importantly emphasizes the power of the mimetic, in which the performance and reality of frontierism come to drive each other.[68] While the figures of the scout and the farmer once dominated this frontier imaginary, I contend that today we are introduced to a new frontiersman of the deep sea: the archaeologist. Like the scout and the farmer before him (and it is historically a *him*, a male gaze of penetration and desire, that populates this space), the archaeologist in the story of the blue frontier has immense power in defining its reality; he is a storyteller, setting the narrative about the way in which this frontier may be integrated with existing infrastructure as well as its possibilities for commodification. Like the scout and the farmer, the archaeologist is seen as an explorer of uncharted regions.

I posit the archaeologist as supplanting the scout and the farmer in order to emphasize the fact that what distinguishes the seafloor as a frontier is not only the physical specificity of its wetness, richness, or volumetric depth but also its relationship to notions of human heritage. To construct the archaeologist as the frontiersman presumes that the frontier can be understood as an archive, and that the archive is itself a frontier. While the scout and the farmer colonized frontier space and used its resources in situ, the archaeologist retrieves those resources and transports them back to land. The resource paradigm of the ocean floor overlaps with the archival paradigm to the extent that both presume preservation of time and extractability-as-recovery. More than a resource frontier, this salvage-extraction dynamic marks the seafloor as what Anna Tsing has called a "salvage frontier, where making, saving, and destroying resources are utterly mixed up, where zones of conservation, production, and resource sacrifice overlap almost fully, and canonical time frames of nature's study, use, and preservation are reversed, conflated, and confused."[69] To better understand the seabed as a salvage frontier, we ought to remember the science of shipwreck salvage and the way in which it already suggests notions of technological betterment, political gain, and the management and circulation of value.

Beyond the exploitation of resources, the seafloor becomes a point of cathexis for ideas about technological, scientific, economic, and political possibility. As Patricia Limerick points out, "sometime in the last century" the spirit of the American frontier "picked itself up and made a definitive relocation—from territorial expansion to technological and commercial expansion."[70] The deep sea offers a convenient space for the hybridization of the old and new frontiers, creating once more a spatial metaphor for domination through the darkness and pressure of the ocean depths, in addition to a technological one in the form of exploration via autonomous underwater vehicles (AUVs), remotely operated vehicles (ROVs), sensors, samplers, cameras, and other novel media devices. The translation of the seafloor through a panoply of new media technologies today lies at the very core of what it means to see the seafloor as a frontier in the first place.

In the war years of 1941–45, oceanographers at Woods Hole Oceanographic Institution were contracted by the U.S. Navy to develop technologies for the photography of shipwrecks.[71]. Maurice Ewing helmed this project, developing a groundbreaking instrument that could be lowered via cables "to any desired depth," making it possible to identify sunken ships and mines. This camera later became the prototype for all subsequent underwater cameras. But the technological innovations that came with nautical archaeology were not just visual. In the 1950s, while working with Jacques Cousteau, Harold Edgerton began developing a device to search for shipwrecks. His invention of side-scan sonar, a towed sonar device, was groundbreaking for its ability to produce a continuous image of the seafloor, and eventually helped locate countless shipwrecks, including the *Titanic* in the 1990s.[72] Today, nautical archaeology makes use of sonar and live-streaming in addition to modeling technologies like 3D photogrammetry, in which three-dimensional measurements are extracted from ordinary photos and video. Such models can be endlessly manipulated and annotated to resemble traditional museum displays.

The various tools and machines constructed to retrieve wrecks are seen as part of a technological frontier initiated by the physical challenge of traversing deep waters. Technologically mediated shipwreck archives, including photogrammetric models, sonar images, and video footage of salvage expeditions, are typically seen as stable architectures in which function is latent. They subscribe to the idea of a "future simple," as Wendy Chun would say, which boils down to an ideal of programmability: the value of an archive or in a database is in its ability to construct a future by learning from the past.[73] Yet technological development is not straightforward—it moves in fits and starts, at varying tempos, and frequently encounters dead ends. As Paul Virilio once noted, "Oceangoing vessels invented the shipwreck."[74] The innovation precipitates the disaster. The continued inevitability of loss at sea, the *future complex* composed of unavoidable yet unpredictable failure—manifest in the form of displaced pasts, premature abortions of technology, and lost futures— is negotiated through constructed versions of pastness that show off innovation and highlight cautionary tales for the future. With CHM, we continue to presume that the past structures the future and that heritage commodities can be integrated into existing managerial regimes.

The Living and Dying Archive

Although digital shipwreck archives produce their own sense of objectivity and consistency, their own "immutable mobiles," as Bruno Latour calls them, digital space differs from object displays in its proliferation of archival forms.[75] The shift in nautical archaeology from artifactual retrieval to telepresence and in situ display has meant that increasingly, the archival process is no longer dependent on retrieval or restoration but rather on acts of audiovisual mediation. More than supporting notions of progress, deep-sea video has produced a distinct vision of what it means to archive the seafloor beyond excavation. In particular, it transforms the seabed into an accessible medium that can itself perform the work of storage, transmission, and display. Organizations like Nautilus Live, a media brand sponsored by famed oceanographer Robert Ballard (who helped discover the *Titanic* wreck) and the Ocean Exploration Trust, have acted as pioneers in ocean telepresence.[76] Many of their excavation videos allow the evolution of the wreck in nature to itself become part of the spectacle. The rediscovery of the USS *Bugara*, for instance, highlights the wreck's process of becoming a reef over the course of forty-six years.[77]

Similar musealizations of man-made reefs have expanded on this notion that the ocean itself is an actor and mediator. This includes artistic projects like the Damien Hirst's *Treasures from the Wreck of the Unbelievable*, a fictional 2017 documentary about the recovery of the sunken treasures of freed slave Cif Amotan II (an anagram for "I Am Fiction"); Ruth Wallen's 2009 *Sea as Sculptress*, a "microphotographic record of the marine life growing on sculptures" placed in the San Francisco Bay; and the MUSA Underwater Museum of Art.[78] It strikes me that digital archives and in situ wreck musealizations have something in common in the way they offer us a view of the continuous sedimentation of history rather than a static picture of the past. At Pearl Harbor, the seafloor is also itself an exhibition space and simultaneously the object of exhibition. This, like the other virtually excavated wrecks and like man-made reefs, is an archive that resists the presumptive activations of matter into linear human history. Rather, these two seemingly contradictory temporalities and narratives coexist. Such a palimpsest of temporalities and stories brings to mind Mentz's composting model of historical change, wherein the past is recycled and multiple presences exist in multiple states of decay at all times.[79] Indeed, this is a composting model of history made literal. That the archive might live and die, rather than be frozen in time, is a feature of the database that transcends our traditional approach to musealization.

There are many archaeologists, media scholars, and indigenous scholars who have pointed out affordances to a living archive. Archaeologists have discussed how the multilayering of information through databases makes it possible both to preserve data and provide access to "the processes that led to the production of that content."[80] This is also

Figure 3. Underwater sculptures at Molinere Underwater Sculpture Park. Photograph via Boris Kasimov, CC.

related to a critique of Western musealization by indigenous peoples. Jeremy Pilcher and Saskia Vermeylen, for instance, fixate on the mechanisms of appropriation, ownership, objectification, and stereotyping of indigenous identity in museum curatorship, arguing that "exhibitions focusing on Indigenous peoples fail to show them as dynamic, living culture." They stress instead a need to provide agency to "social relations that link objects, persons, environments and memories," allowing for a plurality of meanings around heritage objects.[81] This emphasis on social relations better captures the way that heritage is actively enacted among many communities. For example, Karen Ingersoll talks about a "living archive" in relation to Hawaiian indigenous practices like surfing, in which local Kanaka knowledge about the oceans is rooted in history and genealogy and yet also remains organic and evolving.[82]

But while these kinds of digital, multilayered forms of preservation are often treated as novel, there is a continuity between living digital space and living spaces at large that is often ignored. In order to foreground processes of knowledge production in the archive, nautical archaeologists might also include active processes of sedimentation. Including both environmental and digital processes of sedimentation offers a more capacious perspective on heritage and inheritance that exceeds the temporal bounds of extractivism. Julie Michelle Klinger's provocation of "lithosociality" suggests a possibility for a radically human geos that counters assumptions of inhuman minerality that comes to define our sense of extractability.[83] I would argue that by contextualizing the cultural values that are imposed onto the lithos through archaeological excavation and display, as well as vice versa, we can bring forth the humanity latent in geological space as well as the potentialities for relations with the blue archive beyond that of an asymmetrical, proprietary taking.

A lithosocial perspective on the blue archive also allows for a re-thinking of sedimentation itself and its critical affordances. What is sedimentation without extraction? Stephanie LeMenager articulates this most directly in her discussion of the "rivering of time" through sediment, which focuses on sedimentation as the act of settling "into place over time, in ways that might transform the relations of violence bound up in settler colonialism."[84] This form of sedimentation accords both material and semiotic agency to nonhuman processes, and it does not adhere to evolutionary paradigms. To LeMenager, there is a politics to thinking about sedimentation as resistance to extractivism, which, by contrast, operates through a logic of "Do not sediment."[85] I do not see sedimentation and extractivism as operating on binary poles the way LeMenager does, but she makes an important distinction between the two, casting sedimentation as the more capacious term.

What LeMenager, Ingersoll, Kathryn Yusoff, and others have in common in their philosophies is that they decouple the concept of memory from mere storage, placing emphasis on a process of active recall as well as long-term, place-based affinities. This is in sharp contrast to the regulatory regimes governing seafloor artifacts. The original text of UNCLOS tells us that the seabed's artifacts and resources are "inherited" by humanity and that its legacies must be preserved or disposed of for the benefit of mankind as a whole. This assumes, however, that all that is on the seafloor can be excavated, and reifies material excavation as a necessary precursor to valuable or meaningful appraisals of heritage. It appears to exclude possibilities for remembrance that are symbolic or noninvasive, ignoring the diversity of approaches to mourning and memorialization of that which is lost at sea.

Countering this, inheritance requires a conversation about that which cannot be archived or made accessible to the human senses. Thus far, my examples of salvage-extraction have largely come from colonizers, as it is often powerful nations like the United States or China that compete to extract value from resource frontiers. Perspectives from the colonized have provided road maps for how to conceptualize heritage beyond material extraction. Many decolonial scholars have, for instance, fixated on the memorialization of African slaves lost to the Atlantic seabed during the Middle Passage. Writing about the Zong massacre and the regulation of human remains in international waters, Michelle Barron makes a case for attending to "absent bodies," or "those that are ultimately unexcavatable through maritime archaeology or through contemporary legal means, and instead must be figuratively reclaimed."[86] The need to preserve and remember the final resting place of these approximately 1.8 million Africans lost at sea carries with it a powerful international cultural significance—one that transforms the seabed into a sacred landscape that one could argue merits the same kind of care and preservation as a site like Pearl Harbor. Indeed, the *Arizona* stands as perhaps the most powerful memorial at the Pearl Harbor historic sites precisely because it prompts reflection about that which we cannot see but know to be present: 1,177 crewmen killed during a bombing by the Japanese Naval Forces in 1941.[87] Inheritance, then, can be read as an active process of consultation, protection, and enactment—not just the accumulation of artifacts from a prelapsarian past. The archive is always alive.

Like nautical archaeology at large, paying attention to absent bodies is not just an abstract endeavor, but can be recruited directly into the geopolitical regulation of marine extraction. Recently, the Middle Passage has emerged as a potential bulwark against the mining of the deep seabed. One group based at Duke University's marine science and conservation program has attempted to make the case that the "cultural significance of the Atlantic seabed in the context of the trans-Atlantic slave trade" should be "considered alongside the environmental concerns and economic interests associated with deep-sea mining." Mining operations in the area would likely come upon slave-trade shipwrecks and human remains, which the authors argue are deserving of preservation as the "final resting place of the victims of the slave trade." These efforts have been cosigned by prominent scientists working with the ISA such as Cindy Van Dover as a potential avenue for regulating an industry that will undoubtedly damage fragile benthic ecosystems in a near-permanent way (I discuss this further in chapter 3).[88] Practically speaking however, mining contractors continue to hold the funding and the power in this debate, for it is contractors who would be responsible for reporting the presence of Middle Passage wrecks—despite clear conflicts of interest and a strong motive to withhold that information.

Responsible Inheritance

While *inheritance* in its Western configurations tends to signal questions of property and bloodline, the term can also offer space for a broader and more responsible approach to belonging. In "Geologic Life," Yusoff writes, "Inheritance, according to Derrida, requires vigilance about what is inherited and how it is carried forward: 'we inherit it, we must watch over it.'"[89] Indeed, the idea of owning parcels of land and ocean as an enactment of heritage fundamentally contradicts the traditional perspectives of Oceanians at large, who see the Pacific and its "sea of islands" as a vast home "unhindered by boundaries of the kind erected much later by imperial powers."[90] Taking seriously this charge to be responsible for that which we inherit, we must also recognize that we cannot always pick and choose our inheritance. Many human and non-human agencies are at work in dictating how the seabed archive lives and changes. Trawling, surveying, cutting, shipping, and pumping operations transform ecosystems as much as or more than corals do, adding new sedimentary layers to landscapes.

Garbage and pollution, too, are inherited debts. The point of the ISA's legislation is, after all, not simply to determine the belonging of an object but also its disposal; such pollutions affect all corners of the world. Yet many of the by-products of human activity can neither be preserved nor disposed of. The seemingly irreversible cascade of effects produced by oil spills and microplastics, for instance, has often produced more efforts to conceal and to forget than to clean up or remember.[91] Resource pipelines feed back into deep-sea ecosystems in profound ways, from improper sediment discharge carrying free-floating heavy metals for miles through the water column, to light and noise pollution, to other forms of waste that may increase algae production at the surface.[92] Trash and treasure are lively; their temporalities are multiple. While we may treat the seabed as a bulwark of our past and beacon of the future, it affects our lives in ways beyond what is predictable, beyond iPhones, hybrid vehicles, and Chinese shipwrecks.

In this discussion I have shown that salvage and extraction have acted as convergent ideas that mobilize notions of heritage in constricting ways. If salvage, in its technologically mediated and materially mediated formations, introduced the ocean floor as a field of evidence of human histories, it has also provided the justification for the ocean's uses as a resource frontier. Archaeological and legal renditions of the seabed continue to see the environment as a nonrenewable resource to be acted upon by the Enlightenment subject. Videation and educational media perhaps are initiating a shift in that perspective, allowing people to see the seabed as a changing, evolving, or perhaps unruly space in its own right. Yet the salvage-extraction imaginary of the blue frontier continues to produce a powerful, durable, and representationally reproducible imaginary of the deep sea as a space for spectacular accumulation. This "managerial time of the deep" imposes its own ontological framework on the ocean floor, producing categories for the speculative value of seafloor objects as valuable resources, natural obstacles, and cultural artifacts.[93] Progress here is thus a Möbius strip: the human act of reaching the archive suggests its mobilization and uses for the future.

Performing the archival process has in many senses been a performance of modernity, of human evolutions. The language of common heritage as it is used in extractive industry often relies on its broadest connotations, as well as on the preexisting associations around heritage as an inalienable feature of all humanity and an object of political importance. Heritage is a usefully conservative rhetorical gesture that ostensibly captures an entire spectrum of political belief, from indigenous cosmology to, for instance, alt-right insistence on protecting settler monuments as objects of heritage. Heritage might also be thought about alongside the "relational turn" in STS, where the language of relation patently mirrors—without fully acknowledging or maintaining the liberatory politics of—existing indigenous scholarship around honoring relatives and being in relation to land and water. It is evident here that common heritage, and even at times cultural heritage, has been used to justify settler endeavors. As a form of relationality to natural environments, salvage-extraction dominates other frameworks for remembrance, inheritance, or nonlinear perspectives on the past. Although we may still learn from the seabed, we must look to a different, more inclusive framework for this underwater landscape that does not presume that we can merely extract from an archive unproblematically. Instead, all the living things that rely on the seabed inherit it and thus become responsible for its continuation and for its impact on others.

In the deep, *geos* and *bios* are inextricable from one another, as life, death, and decay cycle between one another. As a historical analogy, sedimentation is often divorced from concurrent or adjacent environmental and human agencies in favor of a progressive, linear temporality. However, geological processes of sedimentation do not, in fact, happen in a vacuum. Turbulence precedes and conditions the possibility for a sedimentary archive. In the next chapter I will dig deeper into the turbulence of seabed mediation by examining the history of explosives usage in petroleum surveying, and how this noisy pursuit of fossil fuels is shaping ocean soundscapes and ecosystems today. There we will see once more how media processes, like extraction itself, can be tumultuous and contested.

2

SWIMMERS IN A SONIC PIPELINE
Petroleum Surveys

> One by one the dangers which beset the early navigator have
> been overcome. The chart told him the best course to take
> from one point to another. The mariner's compass enabled him
> to maintain his course when the stars were blotted out by
> clouds . . . with the log and soundings he guarded himself when
> a sight could not be obtained.
>
> —Columbus O'Donnell Iselin, "The Development of the
> Fathometer and Echo Depth Finding" (1932)

A fisherman aims his underwater camera down toward the seabed with
speargun in hand as he peruses a patch of seagrass off the coast of Cyprus.
A distant bang and a sudden lurching of the fish around him cause him
to look up, the camera panning in jerky, oblique angles as if following
a disoriented gaze. For the next minute or so the sounds repeat, occurring at neat, ten-second intervals. There's a deep thudding echo to each
of them—a *kaboomaboomaboom*, or as one commenter describes it, a *chi-
bub bub bub*. Every sonic shockwave seems to shudder through the shoals
of fish in the scene, creating an impression of being trapped underwater.
After a minute or so, the fisherman surfaces, a rush of bubbles filling
the frame.

This unsettling audiovisual scene was filmed and uploaded to You-
Tube by user @potoutus in 2013 under the title, "Hearing Seismic Surveying While Underwater." The caption tells us the sounds are coming
from a Turkish ship *Barbaros*, which the videographer alleges is illegally
breaching the territorial waters of Cyprus to conduct its surveys.[1] After
hours of scouring the web for recordings of seismic surveys, this is one

of few video recordings that I was able to find of a live seismic survey from an underwater perspective. The internet is saturated with industry and educational media that merely animate or model the seismic survey process. Many of these educational videos, such as one on 3D seismic imaging uploaded by a geophysicist, even go so far as to edit in extraneous footage of fetal ultrasounds and animal echolocation to explain sound imaging, effectively minimizing the seismic exploration process.[2] By contrast, this spearfisherman's recording is simple and unedited, and yet haunting as an audiovisual document of the seismic imaging process. Noting how terrifying the sounds are, several of the video's commenters cannot help but imagine how whales and fish might feel, while others speculatively place the booms in a horror film. I marvel at the implication—that the fisherman might be better positioned to empathize with the quarry he hunts than the surveyors on a faraway ship who position themselves as harmless knowledge seekers and ocean explorers.

It is fitting, perhaps, that booms and bangs occupy a space in our collective consciousness that contains both the marking of death, as with the bangs of firearms, and the violent creation of life, as read in the primordial bang itself, the bang that generated the universe—the Big Bang. Frances Dyson calls the Big Bang "a sonic event rather than a sonic continuum. . . . The 'bang' is a noise among an overall noisiness, an identifiable sonic 'thing' or 'event' or even 'object' that stands out, protrudes into materiality, and turns noise—the generalized hum that barely enters language as a category of the sensible—into sound."[3] Indeed, the materiality of bangs both big and small seem suited for the delineation of beginnings and endings. High-amplitude sound waves are experienced as loud volumes that have a manifest materiality—they shock, they immobilize, and they penetrate into rock and earth. J. Martin Daughtry describes violent bangs as the "wartime acoustic sublime: the harsh euphoria of a loud close call with death."[4] But what of the bangs that are a sonic continuum—the bangs where continuity is in fact the point?

In this chapter I trace the transformation of the sonic sea into an audiovisual landscape of petroleum. From a nineteenth-century boom era defined by the bangs of exploration to a new age of offshore drilling, seismic surveys are material-discursive objects that have shaped the environments around the geological structures they seek to capture.

Petroleum seismology as it is discussed here is part of a growing realm of sonic communication of the ocean that includes the clicks and whistles of cetacean echolocation, submarine pings, booms, and other echoic sounds.[5] But survey bangs are standout sounds within this taxonomy, speaking to a different kind of sonic ontology that stands in between noise and signal, life and death.

Like other forms of sonar, hydrological surveys are accomplished by producing a series of high energy acoustic blasts that hit the seafloor and echo back to a set of transducers, which translate sound waves into information about geological structures underneath the surface of the seafloor.[6] Ocean engineer Alfred Keil, the former chief scientist of the U.S. Navy's Underwater Explosion Research Division, made his case for the expansion of this survey work in 1966: "Exploration of ocean resources is naturally dependent on advances in oceanography, but must include techniques for extensive surveys as well as the conduct of such surveys. . . . These surveys must include coverage of those physical characteristics which are controlling for the actual work to be accomplished in the oceans."[7] Keil was just one of many who began to articulate the survey as a distinct genre of mediation. A 1968 position paper on ocean exploration by the National Academy of Engineering distinguishes surveying from research by its "systematic collection programs on regional or world ocean scales."[8] In a similar vein, *Merriam-Webster*'s definition of a survey emphasizes both broadness and precision, citing comprehensive consideration and scrutinization through measurement and data collection.[9] Seismic surveys are meant to be expansive yet highly methodical forms of data collection, used to find the anticlines or upfolds where oil occurs and can be extracted.[10] To achieve adequate breadth, these bursts of sound must be repeated hundreds of times, for days, weeks, and even months. Beyond a single big bang, the use of several consecutive bangs in the simultaneously systematic and broad abstraction of a survey has minimized the sonic event itself as an object of interest as it brings other objects (like oil) into the realm of mattering.

In her analysis of water activism and policymaking in Latin America, anthropologist Andrea Ballestero reminds us that "regimes of knowledge (science), obligation (law), and exchange (economy) constantly shape what we count as material."[11] As we mediate the seafloor, we also compose and delimit a space of reality that validates the presence of certain substances, such as oil, while eliminating others through the calculus of noise and interference. This chapter enacts an oppositional gaze—an attempt to make visible those institutional processes of making matters count, in order to rematerialize that which has been concealed. Problematizing fetishes of technological precision and comprehensive coverage, I seek a return to the noisy, haptic, and explosive underpinnings of seismic surveys, critiquing the erasures of animal life and oceanic materiality that they engender. Beyond wonder-inducing gadgets and technics, beyond increasing efficacy and accuracy, I consider how deep-sea prospecting matters materially—how it affects the space of its interventions.

The most spectacular illustration of this process can be read in the controversies surrounding the ecological impacts of such surveys. The bodies of marine animals, which end up on beaches or, more frequently, rain down to the seafloor in the form of "marine snow," are the forgotten companions to seismic images. Whale casualties in particular, which are occasionally captured in photographs and disseminated by the media, activate a moral outrage that connects activities we may otherwise ignore to the sharp relief of death; yet it is a lust for information capture that produces this violent spectacle in the first place. These nonhuman experiences of acoustic mediation lead me to ask, what are the stakes of producing informatic bodies through vibrations that simultaneously produce carcasses?

While oil has been interpreted in terms of its infrastructures, its cultural legacies, and its environmental implications,[12] there has yet to be a theory of mediation that discusses offshore petroleum extraction in relation to its preceding processes of imaging. I demonstrated in chapter 1 that audiovisual abstraction of the seafloor creates the possibility for extraction by producing it as a space for the taking. This applies to the archaeological media I discussed there, as well as to early narrative media about the ocean—what Nicole Starosielski calls "ocean exploitation films": "Regardless of their subject, the language of battle and hunting pervaded the reception of almost all underwater documentaries. . . . Together these films configured the ocean in terms of its resources."[13] Today, similar tropes of war and exploitation continue to pervade scientific and industrial forms of ocean mediation. They are part of a collective staging for how Westerners relate to the ocean as a frontier. To oil men, a survey is far from merely a perceptual medium; it is a tool of the hunt—a form of extractive mediation that determines not only the taking of resources, but of life itself.

Bringing sound studies scholarship and sonic materialism to bear on the question of seafloor survey practices, I will present a media history of reflection seismology and its vexed relationship to bodies of oil and the bodies of marine animals. First, I explore the binary of noise and signal underwater by examining a series of seafloor technologies, from the "soundfish" to the air gun. This leads to a discussion of the physical effects of sound imaging on marine life, attending to listening practices as impactful sonic events in the ocean. I end with a critique of the way industrial and state actors justify the expansion of seismic surveys on the basis of economic need and national security.

My analysis of seismic surveys below is structured around the concept of the "take." This term comes from legal and professional literature, which defines accidental kills of fish and other marine life as "takes." To me, the use of the word "take" to describe unintended death and injury to marine life merits pause, as it construes animals in the same terms as other substances extracted or "taken" from the ocean, occluding the matter of life and death at stake. Usefully, "take" also has connotations within media production. The word "take" in my headings is thus intended to invoke several materially and symbolically interrelated functions: the imaging technologies that "take" scans of the deep sea, "take" as a signifier of repeated attempts, the physical intake of seabed resources, and in oceanographic terms, the "take" or unintended death and injury to marine life in the deep sea. Each of these takes represent only selective pieces or composites of the ocean floor itself, which contains multitudes that are subsumed, recessed, and overlooked by teleologies of extraction and monetization.

The First Take: Signal and Noise

The history of seafloor sounding begins before acoustics enter the picture. *Sound* derives from the Old French *sonder*, meaning "to sink in, penetrate, pierce," or in nautical terms, "to employ the line and lead, or other appropriate means, in order to ascertain the depth of the sea, a channel, etc., or the nature of the bottom."[14] Early oceanographic research on the deep seabed was primarily carried out through "sounding" techniques that involved lowering rope or wire into the depths and retrieving samples. These techniques reached across disciplines and included industry. In the 1860s, the early days of drilling for oil seeps, submarine petroleum exploration was done by divers who took core samples of the ocean bottom. Later, when the HMS *Challenger* first explored the depths of our oceans from 1872 to 1876, laying the infrastructure for modern oceanography, it conducted 133 bottom dredges and 492 deep-sea soundings. Modern "echo sounding" thus connotes both the sonic and physical valences of the word *sound* in a not altogether unproductive conflation. Stefan Helmreich, for instance, has appropriated the term "sounding" and its association with fathoming as a broad and abstract analytic.[15] Today, forms of echo-based sounding such as seismic imaging would do well to remember this history, as echo sounding implies a sonic materialism. The state-of-the-art tethers that transmit power and signals from control rooms to ROVs in the deep sea today began their genealogy in the lifting of ocean mud from the seafloor to the surface with various types of wire.[16]

It was only with the wartime invention of sonar that ocean "sounding" took on a new, sonic register. At the turn of the twentieth century and the advent of World War I, ocean research transitioned from study within independent disciplines (as with the *Challenger* expedition), toward comprehensive oceanography driven by political and economic aims. As Gary Weir explains, the U-boat menace "provided the catalyst that accelerated American naval oceanographic studies, dramatically altered scientific practice, and profoundly affected the selection of new subjects for investigation."[17] The U.S. Navy had a keen interest in understanding how sound travels through seawater and sediment to meet the U-boat threat, and thus funded much of the research on sound-imaging technologies. Communications scholar John Shiga underscores the emergence of distinctly active, directional forms of detection from this wartime competition in underwater dominance, wherein the transmission of acoustic pulses and the recording of echoes became the principal mode of perceiving marine objects.[18]

The progressive development of bottom sounding technologies and spatially expansive survey techniques dovetails with the classic Shannon and Weaver model of communication, which radically conceptualized communication as the success of signal overcoming noise.[19] In the ocean, this way of thinking about noise as interference and signal as message content is a socially significant categorization that bolstered ideas about the need to facilitate the travel of signal and eliminate the intrusion of noise. Shiga explains, for instance, that the historical division of ocean into signal and noise was achieved through the development of under-water bells, echoes, hydrophones: "Underwater sound was organized to signify symbolically through the association of hazards with the bell sound in nautical culture. Finally, the acoustic field of the ocean was divided into signal (bell ringing) and noise (everything else)."[20] Later, underwater sonar devices diversified the number of signal sounds and meaningful sound signatures, training its listeners "to perceive the ocean through that system of sonic division."[21] This flattening of underwater space was achieved through progressive attempts to eliminate the "noisy" material aspects of seawater so as to facilitate the transmission of desir-able pings and echoes—a teleological pruning of sound that eventually served to erase the deep sea's unruly materiality.

There have been many iterations in the development of such ocean sounding techniques. Notably, in the early 1900s, radio expert and in-ventor Reginald Fessenden developed the groundbreaking Fessenden oscillator while working with the Submarine Signal Company. This was a major advance in underwater communication and detection. Before Fessenden, navigational safety systems consisted of the aforementioned underwater bells located near lighthouses, which could be detected by receivers on ships.[22] Unlike these other methods, which required multiple devices, the oscillator was capable of both emitting underwater sounds and detecting their echoes. Around the same time, hoping to help ships detect icebergs, German physicist Alexander Behm developed a similar echo sounder in response to the *Titanic* disaster in 1912.[23] Later, the use of oscillator technology led to higher-resolution and higher-precision echo sounders. This included the sonic depth finder (SDF) developed by U.S. Navy physicist Harvey Hayes in 1924, which made the first bot-tom profiles of the ocean. Weir notes, "Hayes's SDF turned on a sonic light in a very dark room."[24] At last, sound "began to reveal what years of work with rope and wire sounding lines had only suggested."[25]

For the decades when wire sounding technologies dominated, the deep sea was an unknown and terrifying space for hydrographers and other ocean researchers. The requirement of a mechanical connection to the bottom with traditional sounding techniques was in many ways a burden that exacerbated these unwanted connections to whatever was down there, leading to an underlying desire for observation at a distance. Wireless sounding was the antidote to these fears. When echo depth sounding finally appeared, it was hailed as "a radical and brilliant step in man's mastery of the sea,"[26] emphasizing both the anxiety around the seafloor as an alien, nonhuman space and a desire to control it from a distance. From there, the simultaneous development of better sound sources and more precise methods of sonar-based communication in the ocean was an important factor in the growth in the offshore drilling industry.

The partitioning of sound into adversarial notions of signal and noise could be seen in full effect by the 1930s, when oceanographer Columbus O'Donnell Iselin identified the "afternoon effect," or "the impact that diurnal conditions, specifically the changing temperature of sea water, had on underwater sound transmission."[27] The son of wealthy bankers and a graduate of St. Marks and Harvard, Iselin became a student of Henry Bryant Bigelow, the founding director of Woods Hole Oceanographic Institution (WHOI), at a time when oceanography was seen as a "gentlemanly tradition."[28] Famed oceanographers like Iselin helped to further refigure acoustic transmission in terms of targeting and accuracy through the mathematical elimination of aquatic obtrusions and ocean "noise."[29] Writing on equipment developed by WHOI in 1932, Iselin characterizes ocean life as one instance of interference:

> Unforeseen things are constantly hampering the work of each oceanographic expedition. For example, there are several kinds of marine animals which become wound around the hydrographic wire and stop the messengers. . . . If the submarine "devils" are not interfering with the work, the "devils" of stormy weather are very apt to seize the opportunity to persecute the sleepy oceanographer.[30]

Ostensibly, Iselin is referring to animals such as sea turtles, seals, and dolphins, the same kinds of creatures marked as at risk of being entangled in fishing gear and debris.[31] Iselin's flippant description of such creatures as "submarine devils" lessens the blow of the cruelty of entanglement, as it reduces both ocean and animal life into mere obstructions to the development of underwater communication and imaging.

Following Iselin's research on acoustic transmission, World War II propelled the popularization of echo sounders and other prospecting techniques, as rising demand for oil to support the war effort led to a significant increase in U.S. exploration and discovery of onshore and offshore oil fields. Oil pioneers soon sought out techniques that could provide detailed and continuous, rather than intermittent, information about bottom sediments. Santa Barbara County, home to the earliest offshore oil rigs beginning in the 1890s, was primed to become a leading region for this wartime pursuit of oil. By the mid-1940s, companies like Union Oil Company of California, Signal Oil, Shell Oil, and Macco Corporation were experimenting with new sounding techniques for geophysical surveys in the Santa Barbara Channel. William B. W. Rand, an offshore drilling pioneer in the Santa Barbara area, was one figure involved in the development of modern survey methods. Working first for Shell Oil, then Union Oil Company of California, and later his own survey company, Submarex, Rand advocated for surveying as a way to delineate sedimentary structures through large numbers of observations.[32] He was adamant about the importance of petroleum for national well-being: "Oil companies attempt to provide for our petroleum needs in peace and war, at reasonable prices and at a profit. . . . When the costs of finding and producing offshore domestic oil become substantially higher than such costs for foreign oil delivered locally, then other factors such as security, and national self-sufficiency in oil, must be used to justify the higher cost of domestic oil."[33] This theme of energy security and national self-sufficiency would return again and again to justify the expansion of offshore survey technologies.

Not all survey technologies were high-tech, however. Among the industrial survey methods used during the 1940s is a peculiar device called the "soundfish," developed by the U.S. Navy Electronics Laboratory for geophysical prospecting (Figure 4). The soundfish was a hybrid technology created specifically with the aim of determining seafloor composition. It consisted of a hydrophone encased in a metal container, which could be dragged along the bottom of the seafloor. Frictional noises from the scraping of the metal cylinder on the seafloor would then be picked up by the hydrophone and sent to an amplifier on the towing vessel, providing continuous information about the seafloor. Researchers explain, "Rock makes continuous loud bongs or clangs, sand makes a heavy scraping or rasping noise, and mud makes a quiet swishing noise . . . it is necessary for the observer to train his ear by listening while the equipment is dragged at constant speed over known types of bottom, as determined by grab sampling."[34] Variations of this method with simpler equipment have also been used concurrently, such as the dragging of a hollow metal pipe attached to a wire with an audio amplifier and microphone at the top. As the Navy researchers note, "Some information can even be obtained by listening with the ear near the wire and by feeling the wire with one's fingers. The nature of the tugging and jerking on the wire as well as the noises transmitted up the wire gives some information concerning the bottom character."[35]

This mode of listening and feeling is perhaps striking for its focus on somatosensory perception, which subsumes audition as a sensing paradigm. Hydrophones in this case acted as a proxy for human fingertips, providing detailed transmissions of noise through a highly sensory mode of interpretation. However, the imbrication of listening with feeling

Figure 4. The "soundfish," which consisted of a hydrophone within a metal cylinder, was developed by the U.S. Navy Electronics Laboratory for geophysical prospecting. E. C. LaFond, Robert S. Dietz, and J. A. Knauss, "A Sonic Device for Underwater Sediment Surveys," U.S. Navy Electronics Laboratory, Oceanographic Studies Section, San Diego 52, California, *Journal of Sedimentary Petrology* 20, no. 2 (June 1950): 108.

here reflects the essentially haptic nature of sonic communication itself. Philosopher of technology Don Ihde points to the practice of shaking a closed box to hear the shape of the contents inside as just one example of the way mute objects are given a voice through the percussive exchange between two surfaces.[36] Likewise, the soundfish connects haptics and sonics as it acquires sonic information (via the hydrophone) by first creating the sound (via the dragged cylinder that encases the hydrophone).

Most survey technologies that rely on dragging are limited by imprecision in location data and physical interference, and cannot achieve the level of photographic detail desired by modern oil prospectors. Nevertheless, the soundfish's premise of continuous sounding encapsulates an epistemological desire of the survey, as the ideal of continuous data streams has remained central to the development of seafloor media technologies, even today. The soundfish assumed a subject position in which the oceanographer would be responsible for filtering and interpreting streams of sensory information about the seafloor. Later advancements in deepwater imaging by the military and by shipping cartels followed suit, and focused on improving the shortcomings of dragging methods. These subsequent technologies aimed to elevate signals and reduce noise, cementing a perceived need to manage the deep sea's materiality. Such efforts reflected the increasing conceptualization of the ocean and its animals as forms of interference.

Fundamentally, sound-based survey technologies elided the turbulent middles of transmission through the calculative production of smooth, fluid highways of sound. For echo sounding techniques, unruly noise could refer to things like environmental noise (wind, traffic, marine animals), intrinsic noise (electronic or swell noise specific to the tools being used), reverberations, and "statics," or variable surface conditions that could obscure or change time measurements for the reflections, a crucial element in producing accurate images.[37] In her discussion of undersea cables, Starosielski provides a basis by which we can understand how the elimination of the ocean's materiality extends to the creation of insulated pipelines "designed to transform turbulent ecologies into friction-free surfaces." As she puts it, these infrastructures "produce an internal break in an ecology."[38] I would expand on Starosielski's point to include not only communication industries but also extractive ones. Petroleum surveys likewise construct communication feeds that cut out ocean ecologies. They are pipelines for information that are insulated from both the public and from other ocean sounds—a first step in a value chain that includes the creation of pipelines for oil.

The Second Take: Explosion Seismology

The answer to noise, as it so happens, was the production of more noise. The ocean is a sonic world. Beyond mathematical calculations, an important aspect of amplifying signal involved choosing the right sound source to use for reflection seismology. For all practical purposes, progressive developments of smoother underwater communication depended on loud systems of sonar-based imaging. In water, sound travels four times faster than in air and, unlike sunlight, reaches into great oceanic depths. Extreme bursts of acoustic energy have the ability to travel kilometers from a source and penetrate far into the seafloor. Thus, the higher the energy of the burst and the closer it is to the target, the more accurately surveyors are able to characterize geological structures. The dual development of noise-elimination techniques and noisier technologies can be observed both in early sounding experiments and in modern-day seafloor mediation techniques.

All sound-reflection techniques shared the same basic premise: first, a source transmits vibrations of a particular frequency at pulsed intervals toward the object of interest (the seafloor); then, as the seismic wave travels back to the surface, each reflecting interface bends the raypath. Different materials, whether bedrock, seawater, or piezoelectric crystals, are perturbed differently by the sound waves, changing their ability to penetrate, echo back, or produce electric polarization. The process of energy or signal transformation is called transduction. Piezoelectric crystals, ceramics, composites, or polymers in hydrophone transducers set the limits of what can be "heard" (or transformed from vibrations into an electric signal).[39] At maximum capacity, mechanical stress causes electric polarization, and information propagates through matter.

The first use of seismic imaging technology was the 1921 Vines Branch experiment, where scientists used the phenomenon of echoes off of underground rock to create an image of the space below the seafloor.[40] Over the next decade, reflection seismology grew to become a reliable tool for the location of marine hydrocarbons. In contemporary petroleum seismology, surveyors watch for the particular seismic responses of carbonate rocks, which are source rocks containing over half of the world's hydrocarbon reserves. Their sound signatures are unique thanks to complex pore systems, and are usually measured with ultrasonic transducers.[41] These reflections are then analyzed through various algorithmic filters to produce an image of the subsurface (rock formations below the seafloor).

Eventually, wartime developments led to the appropriation of weapons themselves for the purposes of acoustic communication. That is, explosives, from dynamite to pentaerythritol tetranitrate (PETN, used by the Germans in World War I), were and continue to be an important component of oceanographic surveys. Energetic or explosive materials undergo a rapid chemical reaction resulting in combustion, or the release of heat and gas as the molecular compounds break down. Detonated by a combination of heat and shockwave, high-energy explosives like nitroglycerine release more energy through intramolecular decomposition. They are louder, and thus they retrieve clearer sound images.[42]

Explosion seismology was subsequently popularized by physicist Maurice Ewing, who used TNT to study the continental shelf aboard the U.S. Coast and Geodetic Survey (USC&GS) ship *Oceanographer*.[43] Established in 1807 as the first civilian scientific agency, the USC&GS is the organization responsible for surveying the U.S. coastline and creating nautical charts for the benefit of maritime safety. After Ewing's success in revealing geological characteristics beneath the ocean floor, other USC&GS researchers also began using explosives to make seismic profiles. Electrical engineer and inventor of the Dorsey Fathometer, Herbert Grove Dorsey, chronicled experiments made by the USC&GS ships *Oceanographer* and *Lydonia* in the 1930s. In particular, he generated sound with quarter-pint TNT bombs in order to test hydrophone reception at various distances and depths, measuring refraction and reflection due to changes in temperature, pressure, and salinity. These bomb signal tests, which occurred off the coast of Maryland in 1933 and Santa Barbara in 1934, would lead to the development of more accurate echo sounders.[44]

Dynamite was the original seismic source for surveys because it yielded strong reflection signals and was relatively mobile and compact. But in water, dynamite also had the drawback of producing noisy bubbles of gases, which restricted surveyors to using the explosives in shallow water, further away from the target, so as to minimize the rise of bubbles. During the postwar era, these and other limitations led researchers toward alternatives to black powder as a source of explosion.[45] Safety was perhaps one of the concerns; in 1957, an attempt by oceanographers aboard the USS *Somersworth* to detonate half-pound charges for Mark 3A offensive grenades resulted in catastrophe, killing three people on deck and injuring another four. The *Somersworth* disaster brought to light the dangers of using military explosives as signal sources, particularly without a demolitions expert.[46]

Beyond issues of safety, oceanographers saw a need for more control, "a practical, lightweight, low-frequency, high-intensity sound source, capable of being lowered to actuate at great depths, one which is unaffected by pressure."[47] This goal was eventually accomplished under the watchful eyes (and ears) of oceanographer John Brackett Hersey. A student of explosions pioneer Ewing both before and during the war, Hersey had a background in petroleum exploration, having initially worked with a seismic exploration crew for Phillips Petroleum.[48] In 1947, Hersey was hired to run an underwater acoustics program for WHOI under Iselin, who by then had become the institution's director. Hersey is primarily remembered as a champion for towed instruments, which required sound sources with a greater degree of control. He coupled newer explosive sources with hydrophone techniques to develop continuous seismic profiling (CSP), a technique that consisted of the repetition of echo sounding techniques several times per minute (Figure 5). This introduced a time-based understanding of resolution, in which high resolution equates to the temporal length of the seismic signal.[49] The idea was to create a sub-bottom reflection that would approach a continuous line, an update on the continuity that was first available with towed hydrophone technologies like the soundfish.

By the late 1960s, CSP was widely deployed for offshore oil exploration, and they were used almost as universally as echo sounding. The success of CSP ultimately revolved around increasing signal to noise ratios with controlled, acoustic sources beyond traditional explosives. Some of these alternative sound sources were electrical, creating discharge from spark plugs to generate acoustic signals with broad sound spectra. These could "behave somewhat like explosion, though much weaker."[50] For instance, the "Boomer" was an eddy current generator developed by Harold Edgerton at the Massachusetts Institute of Technology. Edgerton's 1964 paper introducing the device explained that the Boomer could "be triggered often at an accurately controlled time, thus enabling the user to correlate the results of many operations on a tape or chart." It also reduced problems with noise via a digital correlation technique, and provided a large energy pulse of lower-frequency sound.[51] Parallel developments also emerged from the laboratories of Socony Mobil Company and Lamont Geological Observatory of Columbia University. At Lamont, small charges of TNT were deployed for "several nearly continuous world circling profiles."[52] Sometimes, CSP even involved the simultaneous use of two different kinds of sound sources.[53]

But by far the most significant development occurred in the 1970s, when Lamont and manufacturers Bolt Technologies and Texas Instruments pioneered the use of air guns, which used blasts of pressurized air as a sound source. By the mid-1970s over 50 percent of marine seismic

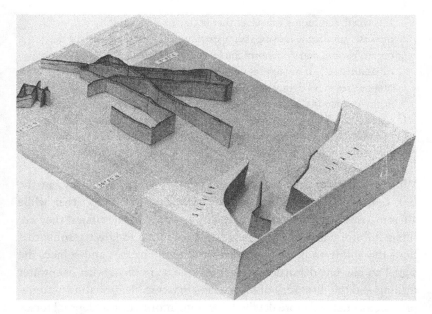

Figure 5. A model of a CSP survey in the Tyrrhenian Sea, Woods Hole Oceanographic Institution. John Brackett Hersey, "Continuous Reflection Profiling," in *The Earth beneath the Sea*, ed. M. N. Hill (London: John Wiley & Sons, 1963), 61.

surveys relied on air guns.[54] Like dynamite, air guns create extraneous bubbles, but this problem is remedied by the use of multiple, consecutive bangs. Specifically, differently sized air guns are fired simultaneously in an array so that their pulses sum together and can be "tuned" to minimize the size of bubble pulses.[55] The echoes produced from these bursts are then recorded by up to three thousand hydrophones that stream from the ship.[56] Today, air guns are typically towed from survey ships and arranged in a square array below the waterline, where they fire five or six times a minute at 200 to 240 decibels. When translated from an aqueous context to air, this is the equivalent of 140 to 180 decibels, which approaches the threshold for human pain and long-term hearing damage.

Driven primarily by the anxiety about petroleum and hard mineral interests, Hersey and his colleagues in underwater sound had a distinctly extractive understanding of the seafloor and a realist perspective on the institutional funding of marine science. In a 1971 speech, Hersey stressed the importance of industry in funding marine science:

> It is worth reminding ourselves that both petroleum and hard mineral interests are already moving their experimental operations into the deep ocean. Various departments of the federal government need deep ocean capabilities. . . . If my figures are not woefully dated petroleum investment at sea at all depths is a few billion per year, and the federal government spends slightly over half a billion on what is classed as marine science. . . . Each must make his own counsel regarding this influence, but there seems little doubt that the wealth and the understanding of the oceans will need to be exploited increasingly in years to come.[57]

With a foot in both the scientific and industrial worlds, Hersey saw the link between surveying and drilling. He also understood that while air-gun arrays towed from surface ships can reveal the basics, they are often still noisy and low in resolution, generating long wavefronts that limit the ability to determine small structural changes and subject the signal to current distortions. This became a problem with deepwater drilling and the discovery of oil reserves inaccessible in shallow water— prospectors needed more detailed, accurate information at depth. Hersey thus realized a need for bottom profiles with "near-photographic detail" beyond what existing echo sounders, which he thought were "meager and rather clumsy," could do.

Responding to the growing deepwater drilling industry and problems with increased depth, the mid-1970s saw the development of ocean-bottom seismographs, which could return more accurate location information with better signal-to-noise ratios.[58] Bottom technologies sought to offer more direct forms of penetration with precise energy points by placing both the source and the receivers (hydrophones) on the seafloor instead of towing them behind a ship.[59] This allowed surveyors to retrieve precise wave velocities within thick sediment columns.[60] By 1975, deepwater drilling was in full swing, exploiting depths of five hundred to one thousand feet or more. Bottom technologies continued to grow in their scale and sophistication as drilling interests moved toward deeper waters like those in the Gulf.[61]

Jim Broda, a researcher at WHOI, led a project in 1990 called the Near Ocean Bottom Explosive Launcher (NOBEL), the first imaging system to detonate multiple high-explosive charges at the seafloor. I caught up with Broda at WHOI, where he explained his experience with bottom seismographs in relation to his own invention:

> We'd make a bomb, literally, strap together six boxes, eight boxes, or up to twelve hundred pounds of TNT, light a fuse on the back deck of the ship and throw it in the ocean. . . . TNT was the first phase, but I ended up dealing with some of the most extraordinary high energy, insensitive yet energetic materials, warhead grade. The results we got were extraordinary. . . . It'd be like, I used to study the moon with binoculars in a lounge chair in my backyard and now I'm standing there. That's the leap in resolution.[62]

Broda's use of a visual metaphor to describe this effect highlights the way in which sonic information is rendered and understood culturally as akin to visual media forms. Broda and other marine geologists talk about sedimentary formations in terms of resolution and clarity, equating higher energy release to higher image fidelity. The bigger the bomb, the better the picture. Today, as they probe seafloor muds for clues of oil, industry actors construe both acts of sounding and acts of drilling as similar forms of dimensional or deductive listening, in which the production of sound through physical contact or impact is used to construct spatial information in the absence of vision.[63] For example, in 2015 the American Petroleum Institute (API) released a report describing the oil and

Figure 6. The Near Ocean Bottom Explosive Launcher at Woods Hole Oceanographic Institution. Photograph by author, 2019.

gas industry's desire to capitalize on unexplored reserves from the Outer Continental Shelf (OCS).[64] The report contains contradictory statements about the nature of drilling and imaging. The first puts the onus of knowledge production on imaging: "If Congress permits the use of state-of-the-art seismic surveying technology in largely unexplored areas of the Atlantic OCS, we may discover an even greater abundance of oil and natural gas." The second, meanwhile, reverts back to drilling: "If you can't drill for oil and natural gas, you can't know how much you have."[65] Here, both drilling and surveying serve to tell us "how much you have." The offshore oil industry thus justifies its expansion through a conflation between drilling and surveying as performing the same work of knowledge production. Given the collapse between these two practices, soundings are aptly described as extractive mediations—media practices whose fluid collection of acoustic signals preempts and mirrors the production of resource pipelines. The search for petroleum links the pursuit of knowledge and the pursuit of economic wealth.

Reflection seismology, and the kind of dimensional listening that accompanies it, is an "audile technique," a sonic practice that is informed by and affects social contexts beyond its mere object of study. Jonathan Sterne unpacks the idea of audile technique through the example of auscultation, wherein the medical field's association of listening with knowledge and skill transforms audition into a mode of power and a marker of middle-class identity.[66] Likewise, both informatic and material, seismic surveys in the ocean corporealize and operationalize landscapes according to existing social hierarchies. The event of seismology constitutes, in a sense, the event of a birth; it is the emergence of the oil reserve as a measurable body of information, energy, and capital. The result is both the creation of new meanings and matters and the elision of others. But what of those properties of sound that we discard—its material existence beyond signal propagation? In the next section I explore relationships between sound, water, and life, attending to how seismic imaging is experienced by marine animals.

The Third Take: Ecological Costs

Sound has a long history of association with liquidity. The metaphors abound: music hits us like a waterfall, it flows like a river, it immerses like the ocean, it crashes like a wave. In literature, in song, and in scholarly work, sound and water are often woven together as two ur-liquids, reflecting each other in their mutually shared fluidity, their transmutability, and their undulating, wavelike qualities. Veit Erlmann is one such scholar who has explored the metaphysical and physical resemblance between sound and water within a historical context, as the foaming waters of the Rhine Falls inspired a convergence between the aural and the erotic for early Romantic writers like Wilhelm Heinse. In the late 1700s and early 1800s, water was understood to be a vital fluid, or *fluidum vitale*, symbolizing Romantic subjectivity in the fusion of noise with euphoria.[67] The relationship between sound and water exists even on an etymological level. Erlmann notes that the German verb *rauschen* "signifies the sound made by things as rivers, brooks, and leaves, and thus might be translated as 'rustling,'" while the noun *Rauschen* denotes white noise. In addition, the word *noise* itself is theorized to be related to either the Latin word *nausea*, referring to seasickness, or the Latin *noxia*, "hurting, injury, damage."[68]

Echoing Heinse two centuries later, Helmreich revisits this association in his book *Sounding the Limits of Life*, in which he ruminates on the shared ontologies of water and sound as traveling entities. Helmreich discusses transduction as a quality that builds on notions of sonic movement and flow, connoting crossings through space, time, and life itself.[69] While Erlmann's text chronicles the early Romantic vital fluids, and Helmreich to some extent echoes these vitalist assessments of sound and seawater, formalized understandings of transduction today have a tendency to frame sound in empirical terms. Where once the affective experience of sensory saturation produced notions of the sublime, today these thresholds are action potentials, discussed in terms of their functionality for humans. Even as the metaphor remains, we have begun to move away from experiential contexts for thinking about transduction, instead operationalizing it in the context of media operations and the acquisition of signals. Yet here, while examining the extractive mediations of explosion seismology, we can pause to remember those earlier associations between sound and fluid. Rather than contain ourselves to the communicative capacities of sound waves, we can reencounter underwater noise as deluge, as affective saturation.

Important scholarship in acoustic ecology, anthropology, and history has pushed back against the noise/signal binary as a one-way transmissive model, finding ways of validating noise itself as a cultural object.[70] For instance, philosopher Michel Serres discusses noise in terms of the figure of the parasite, drawing attention to the vitality of process, propagation, and mediation: "In the beginning was the noise."[71] Building on these works, I contest this division of ocean into signal and noise through the bang, that object which is both noise and signal, which prompts immersive feeling while simultaneously communicating information. And bangs, importantly, lead us to nonhuman formations and the differential experience of noise in the ocean by whales. Indeed, it is with a consideration of whale hearing, and with the deafening of marine inhabitants, that I revisit sound as a haptic force that spills over the bounds of information and signal into nausea, overload, and noise. While seismic surveys affect swimmers of all shapes and sizes, I want to take a closer look at whales in particular, which have been part of an extractive media assemblage that has dictated the course of human ocean exploration for over a century. Both abjected and revered by humans, these cetaceans have become the proverbial canaries in the coal mine, or to put it in more precise terms, swimmers in a sonic pipeline.

We often think of whales as nomads, traveling in bands around the world and migrating with the seasons. But many of these gentle giants prefer the sedentary life. In the waters of the Gulf of Mexico we find one such example in the Rice's whale, a recently discovered, distinct species of baleen whale found nowhere else on Earth. *Balaenoptera ricei* is a critically endangered species, with roughly thirty to fifty individuals remaining in the Gulf of Mexico, primarily in the De Soto Canyon, sixty miles offshore from Pensacola, Florida. The Gulf region has proved a perilous home for marine life since the advent of the offshore drilling industry and its accompanying externalities. The 2010 Deepwater Horizon oil spill killed roughly 22 percent of this group, and the expansion of seismic oil surveys has steadily whittled down the population even further.[72] Today, the waters of the Gulf are among the loudest and most polluted in the world. Every summer, a regional dead zone, a product of nutrient pollution from agricultural and urban areas, manifests, and it is steadily expanding. Meanwhile, De Soto Canyon is a busy commercial area, constellated by offshore drilling platforms and cargo ships. Rice's whales are deep divers, feeding both at the surface and at the bottom. But while this might limit visual access to the whales, it is no barrier to the deeply penetrating bangs of oil surveys, which are estimated to harass this struggling community thirteen thousand times per year. Rice's whales are not alone; also imperiled by seismic surveys are North Atlantic right whales, which make their homes nearby off the southeastern U.S. coast. And with a population now under 350 individuals, they have been listed as endangered since 1970.[73]

Impacts of anthropogenic noise on whales have been very well documented. According to a 2009 study by a Scripps Institution of Oceanography researcher, ambient anthropogenic noise has been doubling in intensity every decade for over sixty years.[74] In 2012, scientists measuring ambient noise levels and tracking the calls of North Atlantic right whales have estimated that right whales have lost 63–67 percent of their traditional communication space due to man-made noise.[75] Blue whales, fin whales, gray whales, right whales, and humpbacks sing complex, locally specific songs to navigate and communicate with one another in a manner resembling dialects, constituting, as Margaret Grebowicz notes, "the largest communication network for any animals, with the exception of humans."[76] For whales, hearing air guns is roughly like hearing gunshots in sequence or being immersed in an extremely loud rock concert. Additional noise in the ocean from shipping and sonar affects migration, mating, and other social behaviors.[77] Many cetacean breeding grounds, for instance, including those of humpbacks and right whales, occur in the warm coastal waters of the South Pacific.[78] Their calves, however, are easily stimulated by noise, and thus increases in noise around these coastal areas means that key sanctuaries are being lost. The disruption to these nonhuman networks of communication puts the anthropocentric characterization of surveys in sharp relief, revealing the many ways in which our ambition to clarify one type of communication signal can interfere with the signals of others.

Whales and other cetaceans perceive the world through large auditory organs that can determine sizes, shapes, speeds, and textures of objects. Unlike human beings, they hear just as well at depth as they do on the surface.[79] Cetaceans, who navigate, hunt, and form social groupings primarily through echolocation, hear in a haptic way. In fact, toothed whales do not hear through an eardrum and transduction through the middle ear like humans do, but rather through the fatty tissues in their head and jaws, which connects sound vibrations to their inner ear via an acoustic funnel.[80] This is a hapticity that fuses not the eye and the hand, as Gilles Deleuze and Félix Guattari would say, but the ear and the hand; it is the discovery of touching within the hearing function.[81] A similar fusion of haptics with sonics in the hearing function is perhaps more easily grasped in the concept of something like bone conduction, in which vibrations trigger the inner ear via vibrations in the jaw. Certain types of hearing aids use this principle to bypass the eardrums completely. This is also the reason why one's voice sounds different and perhaps fuller in one's own head than it does to others at a distance. Spatial distance between the source and reception of a haptic sound, as Deleuze, Guattari, and Colin Milburn explain, does not act to separate, but rather serves as a medium of passage.[82]

Earlier, I posited that seismic surveys are a particular audile technique that involves dimensional listening, or listening for three-dimensional spatial information. This term comes from radio historian Susan Douglas, who distinguishes between modes such as "dimensional listening," "informational listening," and "associational listening" (an emotional mode of listening), positing that while different modes may sometimes converge, they are tied to specific historical, situational, and medial contexts.[83] These discrete typologies of listening tend to encounter limits, however, when one considers the broader tapestry of listening in the animal and inhuman world. In particular, the plasticity inherent in crosssensory or synesthetic perception challenges and decenters the listening that emerges from a framework of human audile technique.

Speaking to the physical processes of hearing, sound theorists such as Steve Goodman and Daughtry position sound within the framework of vibrations, accounting for aspects of sound that exceed the disembodied ideal ear. To them, vibrations instead push us to consider sound as a phenomenon that is simultaneously haptic, sonic, and affective. As Goodman puts it, "Sonic culture, thus situated, renders the urban audiosocial as a system of speeds and channels, dense pressure packets, vortices of attraction, basins of acoustic immersion and abrasion, vibratory and turbulent: a whole cartography of sonic force."[84] Vibrations, which radiate outward, implicate several bodies and surfaces at once, pointing us to "earth-bound circuits," a structural and processual mesh of geology, biology, and technology.[85] Michael Gallagher, Anja Kanngieser, and Jonathan Prior describe listening to landscapes as a vibrational exchange, arguing that sounding a landscape positions human subjectivity at the margins of "many resounding bodies."[86] Given the assemblage of actors involved in acts of sounding, it makes sense to depart from thinking in terms of discrete objects and move toward what Alfred North Whitehead termed "superjects," "where everything—even a stone, as Whitehead would say—counts as an experiencing subject."[87] The materiality of noise can thus orient its readers toward the intimate mediatory capacities of multiple bodies, including bodies of water, land, and animals.

Bangs also raise concerns about the thresholds at which acoustic vibrations become a violent physical force. In fact, there are several instances in which the physical impacts of loud underwater sound have led to mass whale strandings. Images of these strandings have circulated easily in the media and are quickly seen as tragedies. The heart-wrenching story of sonar-based technologies leading to whale beachings was most notably recounted by journalist Joshua Horowitz in his 2014 book, *War of the Whales*. Horowitz focuses on the accusation that Navy sonar is harming cetaceans and causing mass whale strandings, an issue that exploded into public consciousness in March 2000 with the beaching of seventeen whales (including several Cuvier's beaked whales, Blainville's beaked whales, Minke whales, and a spotted dolphin) on the shores of the Bahamas.[88] It was an extraordinary and dreadful event. Only two other mass strandings of beaked whales had been witnessed since 1864, and this one went down in the books as one of the largest multispecies whale strandings ever recorded.

Although charges of environmental sonic violence were initially met with denial, a retrospective analysis of a string of mass stranding incidents led to a series of lawsuits against the U.S. Navy and the National Marine Fisheries Service for the deployment of low-frequency active (LFA) sonar, typically used to detect objects like submarines over long distances of three hundred miles or more.[89] On August 7, 2002, the National Resources Defense Council (NRDC) filed a landmark lawsuit against the Navy and the National Marine Fisheries Service regarding the deployment of LFA. Prior to the lawsuit, LFA was permitted in 75 percent of the world's oceans.[90] News reporting on this controversy describes sonar as both rocket bursts ("as loud as a Saturn V rocket"), which suggests a kind of discrete targeting, and floodlights that connote dispersion effects.[91] In a Joint Interim Report published by NOAA after the event, it was found that the whales "experienced some sort of acoustic or impulse trauma that led to their stranding and subsequent death." The report continues, "The most significant findings, which were found in the two freshest specimens, consisted of bilateral intracochlear and unilateral temporal region subarachnoid hemorrhage with blood clots bilaterally in the lateral ventricles."[92] Simply put, the loud Navy mid-range frequency sonar caused the whales to hemorrhage, which led to cascading physical debilitation, including overheating, physiological shock, cardiovascular collapse, and severe compromise of hearing and navigational abilities, resulting in stranding.[93] There are marine mammal researchers today who still see Navy sonar as a bigger threat to whale populations than seismic surveys.[94]

While noisy waters can affect whole ecosystems, cetacean takes have clearly gained a particular notoriety. Since the Cold War era of submarine and cetacean science, Americans in particular have embraced cetaceans as intelligent beings, distinct from other animals because of their perceived proximity to humanity.[95] As fields such as animal studies, critical race studies, and the environmental humanities have noted, this exceptionalism is a double-edged sword. By ascribing human qualities to whales, we allow them symbolic status in our world, while reifying animality as the grounds for dehumanization.[96] Such a view fundamentally ignores multispecies affinities and frameworks for justice, extending the privileges of humanity to others instead of recognizing the animal as inherently a part of humankind. On a material level, however, we cannot separate our treatment of other sea creatures from that of whales, since it is not just whales who are imperiled by anthropogenic sound in the ocean.

Some of the early controversies around seismic imaging experiments by petroleum companies focused on their impacts to a wider range of marine species. For instance, in 1948, offshore seismograph surveys near the Santa Barbara shoreline led to protests and an official complaint from the county, which elaborates, "Said blastings have been killing a great quantity of fish and other sea life, along said coastline, and have endangered the lives and property of the residents of said County and said blastings have further interrupted the peaceful enjoyment of the beaches and parks by the people of said County of Santa Barbara."[97] A resolution was subsequently passed, requesting that state agencies, including the Fish and Game Commission, "take appropriate action to protect the County of Santa Barbara and particularly the beaches and parks from off-shore blasting operations."[98] A cross-comparison with 1948 archival documents from Union Oil reveals a more precise picture of how these surveys were deployed (see Figure 7). From March to October 1948, geophones and dynamite charges were suspended from surface floats four to six miles offshore as well as near the middle of the Santa Barbara channel, south of the city. Recording boats followed "shot boats," which dropped floated charges four to five feet below the surface. The crews used jetted charges—explosives buried ten to fifteen feet below the sea-floor by water jets—in order to minimize fish kill, yet even with 214 jet shots, the Union Oil report estimated that the weight of fish killed by the surveys during this time was roughly twenty-five tons.[99]

Meanwhile, while billed as "more environmentally friendly than explosives,"[100] modern air-gun surveys, which release blasts of compressed air, create pressure waves that can penetrate several hundred kilometers into the ocean floor and have a wide range of negative impacts on whales, fish, and invertebrates.[101] In close proximity, the effects are extreme. U.S. Army engineers have detailed the manner in which seismic blasting creates a physical, pre-acoustic shock that can result in everything from animal suffocation to organ damage. This happens because physical shock travels faster than the acoustic velocity of an explosive.[102] Animals near a blast can experience immediate hearing impairment, while fish eggs and larvae can be killed by the explosive pulses. Tension waves generated from explosions have been shown to be particularly harmful to fish with gas-filled swim bladders. More recently, a 2017 study on air guns showed adverse impacts to zooplankton, causing two- to threefold increases in dead adult and larval zooplankton and catastrophic death to larval krill in the air-gun passage.[103] Zooplankton are crucial players in the ocean food chain; disruption to their population creates concerns for fish and marine mammals alike.

Mitigation and regulation of sound in the ocean has often been a matter of creating cartographic maps of the ocean that delineate marine protected areas.[104] For instance, Amy Propen offers a detailed discussion

Figure 7. Dynamite blast during marine seismograph survey by Union Oil
Company of California, 1948. Image from William Whitehall Rand Papers,
SBHC Mss 46, Department of Special Collections, Davidson Library,
University of California, Santa Barbara.

of the role that mapping and the invocation of boundaries between ocean biomes played in resolving the NRDC case. For both sides, the cartographic representations relegate animal takings to specific regions. Propen sees maps and countermaps as "visual-material rhetorics" that freeze fluid regions in both time and space for a particular context. Their durability, reproducibility, and strategic use against each other directly influence policy and corporeal experience in the areas represented.[105] Ignoring the ocean's fluidity, high-resolution images and maps of the sea are treated as synonymous with the sea itself, exposing the ocean to a form of prehensive violence in which destruction is justified through its containment in abstract spatial and temporal terms. For example, current seismic activities are made to stop when whales are spotted up to fifty-six miles from the blast site, but this likely means little in an environment where sound can travel as far as two thousand miles. As whale researcher Scott Kraus puts it, "All of these seismic activities are a little bit of a lipstick on a pig. That is to say they will prevent immediate mortality if a whale gets so close that it's going to get blown up."[106]

That said, not all marine conservation researchers have such an adverse relationship to the petroleum industries. Elizabeth Burgess, a marine mammal hormone researcher at the New England Aquarium, started out in 2005 as a contracted marine mammal observer for the New Zealand's Crown Minerals—in fact, she was the very first dedicated marine mammal observer for seismic surveys in the country. Burgess explained to me that her job was to spot whales (a process that was, at the time, simply done by eye and occasionally through night vision goggles) and notify surveyors when to halt operations to avoid harming nearby whales. She remembers this as a pivotal experience that led her to realize a need for more research on the impacts of anthropogenic activity and noise on whales. Not unlike the spearfisherman at the beginning of this chapter, Burgess's proximity to marine life through extractive industry enabled her to develop a sense of care for those whales. Today, better techniques such as audio observation are used, and observer programs on petroleum survey ships have proliferated.[107]

For these and other reasons, reactions to the impacts of anthropogenic sound on whales fall short of banning sonic technologies, and instead recommend case by case consideration and mitigation strategies. A 2002 report by the Scientific Committee on Antarctic Research

Ad Hoc Group on the "Impact of Marine Acoustic Technology on the Antarctic Environment" recommends uses of minimum source level, careful laying of survey lines, avoidance of repeat surveying of an area in consecutive years, and the use of "'soft starts' whereby power is increased gradually over periods of 20 minutes or more."[108] The idea of soft starts is essentially an animal warning that resembles older mitigation strategies. In a 1997 report on how to mitigate environmental effects of underwater blasts, for instance, engineers Thomas Keevin and Gregory Hempen recommend a combination of helicopter aerial surveys and smaller blasts from shell crackers or "seal bombs," which would ideally "'scare' marine mammals from the blast zone prior to detonating the large explosion."[109] As noise becomes the solution to noise, the induction of fear through sound becomes a matter for animal survival.

This portrayal of mild sonic bangs as a form of risk mediation fits neatly into the calculus of extractive mediation, centering the perspective of a knowledge-seeking human researcher as responsible for discerning noise from signal and minimizing external impact. It largely ignores or deems negligible the production of fear, anxiety, and confusion in nonhumans. However, evidence of these effects for marine species abound: whales change their vocal behaviors around seismic surveys, either calling more frequently or ending their singing around operations.[110] Meanwhile, white whales were found with increased norepinephrine, epinephrine, and dopamine levels after seismic air-gun exposures, while bottlenose dolphins have shown increases in aldosterone, all indicating stress.[111] From an industry perspective, where life and death is the only binary given consideration, injury or affective violence does not matter at all. Seawater, harmful noise, and ocean currents also cease to matter in a context where their materiality is eliminated in calculations for sound propagation and in the building of sonic pipelines in the deep sea.

Building on Goodman and Daughtry's framework of vibrational sound, we may consider here an expanded sense of the belliphonic beyond human pain, turning instead to its affective and physical action on nonhuman subjects. Shell crackers and seal bombs bear some resemblance to the deployment of long-range acoustic devices used to control bodies and crowds in the event of a protest in the human world. Such sounds extend to pre- and parasonic realms, causing temporary debilitation, panic, and confusion within a group. This "unsound" is a becoming tactile of frequencies that are both abrasive and affective. While warning blasts might reduce numbers of seismic casualties, they are predicated on incapacitation through acoustic shock, resulting in the inducement of flight responses.

Ultimately, cetacean death, while affectively charged and mobilizing when occasionally photographed on our beaches, remains elusive in its professional and industrial contexts. After the 2002 LFA lawsuit made sonic violence in the ocean a matter for public scrutiny, the Pentagon chose to lobby for numerous exemptions to the Marine Mammal Protection and Endangered Species Act as a means of creating justifications for continuations of harmful sonic activity. A list of active and expired military "Incidental Take Authorizations" for accidental animal killings is available on the NOAA Fisheries website and includes LFA surveillance, mine reconnaissance, and acoustic technology experiments.[112] In 2018 the Trump administration allocated five new Incidental Harassment Authorizations (a type of incidental take) for the National Marine Fisheries Service, allowing seismic air-gun testing in the Atlantic by oil and gas companies, posing substantial challenges to previously hard-won campaigns to minimize such sonic violence in these waters. The move was sharply criticized by scientists, coastal businesses, communities, lawmakers, and fishermen who argue that such blasts could be detrimental to ocean life, particularly to the North Atlantic right whale, which could be driven from endangerment into extinction.[113]

The oil and gas industry continues to avoid acknowledgment of the harmful effects of seismic surveillance in publicly available press releases, websites, and reports. Despite ample evidence and even Navy acknowledgment of the harmful effects of anthropogenic sound, internal oil industry reports and press releases continue to narrativize seismic imaging as a harmless form of surveillance. Groups like the API and Petroleum Exploration and Production Association of New Zealand (PEPANZ), now rebranded as Energy Resources Aotearoa to incorporate the Māori-language name for New Zealand, issue blanket denials, insisting that surveying, the "first step" in oil extraction, is below a threshold of harm to the environment. To underscore its innocuous nature, the API and other geophysical organizations frequently call these imaging processes "ultrasounds of the earth," infusing them with a maternalistic imaginary.[114]

On a physical level, ultrasounds do operate under the same principle of echoic imaging, but the two sonic technologies are on different orders of magnitude and can hardly be compared in regards to actual physical safety. The invocation of a diagnostic medical technology also seems to appeal to a notion of seismic imaging as risk-minimizing technology.[115] Even if the analogy were to hold, feminist work on ultrasound has discussed the complicity of fetal imaging in anti-abortion political messages, making the point that ultrasound is not an innocent window into the fetus, as it helps to produce what it images and to some extent can dictate political and cultural debate.[116] In reflection seismology, what is coordinated, captured, and controlled is information about oil reserves—the object that, like a fetus, is nurtured and monitored from afar. In this vein, seismic images are social technologies that likewise shape and define what they seek to capture.

The reassuring (yet often misleading) gesture by oil and gas websites toward safety and environmental mitigation is usually followed directly by an insistence that whatever is being done to the ocean by the industry is essential to the well-being of nations. For instance, in its justification for offshore oil, the API states that "in order to ensure our energy security and create economic growth it is vital that we take advantage of all our energy resources, including those safely developed in American waters."[117] Prior to its rebrand as Energy Resources Aotearoa, PEPANZ also included a page on seismic surveys insisting that it performs a necessary role of "contributing billions to our national economy and providing energy security for kiwi households."[118] Discourses of security have always been motivating factors for the production of logistical media. As Ned Rossiter explains, this spans as far back as the Cold War, when there was a symbiosis between the rise of Fordism, security discourse, and the political threat of contingency and destruction.[119] For the petroleum world, the idea of "energy security" blends security discourse with the specter of resource depletion that plagues the Anthropocene. Imre Szeman calls this the logic of *strategic realism:* "At the heart of strategic realism stands the blunt need for nations to protect themselves from energy disruptions by securing and maintaining steady and predictable access to oil." He continues, "Strategic realism sees the disaster of oil as a problem primarily for the way in which nations preserve or enhance their political status."[120]

However, there is an inherent necropolitical implication to this rhetoric of exemption and political authorization, wherein large regions of animal habitation are suspended as zones of exception, deemed to operate in the service of civilization or national vitality. Takes become sacrificial acts, necessary evils that are justified through the simultaneous validation of oil as a life-giving, nation-sustaining substance. Achille Mbembe explains, "Sovereignty means the capacity to define who matters and who does not, who is *disposable* and who is not."[121] Necropolitics, as a question of deciding what matters, can take mattering in both its literal and figurative forms. Survey work determines which objects are allowed a material body, which objects are disposable, and what can or cannot be taken. The imaging process becomes an occupation of a geographical area both physically and visually, territorializing the deep sea by allowing sovereign control over a region from a distance.

The Final Take

As land-based resources shrink, oceanic surveys and resource prospecting are becoming central to the maintenance of an industrialized, and now digitalized, society. It would be dangerously naive to dismiss this industrial exploration itself as harmless. As scientists already implicitly acknowledge, the little information that is produced about the seafloor tends to get refitted for the purposes of attracting investors. It becomes imperative to question not only the authority of these data images themselves but also the imaging process and its potential to cause disturbances in secluded areas of the ocean before extraction even begins.[122]

Standing on the shore, stepping between the bits of tar speckling the beaches by UC Santa Barbara, there is an intuitive, affective connection to oil here—a kind of casual, emplaced petroculture. From the shallows, our ways of seeing oil, feeling oil, and smelling oil connect us to extractive industries on a bodily level. Yet on the level of hearing and sounding, that same kind of embodied intimacy with oil industries remains conspicuously absent. Seismic blasting is not the sort of thing human beachgoers complain about on a regular basis. But they are becoming an increasingly common experience underwater, disturbing the peace for cetacean communities. Sterne talks about a tendency to associate sound with interiority, subjectivity, and proximity, but, as I hope to have shown, the sounding of oil in marine contexts presumes notions of objectivity, exteriority, and distance, connecting us to this oily, watery world in caustic ways.[123] As anthropogenic noise in the ocean worsens and offshore oil extraction moves to ever deeper and more remote waters, focusing on the easy, accessible kind of environmental awareness located in sporadic and spectacular events like oil spills is not enough. Although we may be situated at a distance, there are high stakes to acknowledging the fundamentally necropolitical operations of offshore seismic imaging.

In this chapter I have characterized the bangs of petroleum surveys as a haptic and sonic force, offering a perspective on the relationship between hearing and feeling. As Tim Ingold has stated, "the ways of acting in the environment are also ways of perceiving it."[124] This fundamentally troubles the epistemological premise of survey, which emphasizes objective, continuous perception. Yet here the uncertainty principle is also itself a certainty—the very act of observation affects that which is being observed. While our dispositions toward sonic images is typically trained on the end point of informational filtering of signal from noise, thinking about the excesses of sound in an underwater environment forces us to anneal resources to their oceanic substrates, to their animal inhabitants, and to our technologies.[125]

In the following chapters I will continue to explore relationships between extractive mediations and the multispecies communities they impact. I revisit our relationships to whales in particular in chapter 4, as they are recruited into large-scale assemblages for remote ocean observation. These chapters further demonstrate that recognizing moments when mediation itself becomes extractive grounds human responsibility in a world that exceeds our own values and perceptual limits. What else can we encounter in the space between noise and signal?

3

PERILOUS PLUMES
Mining Waste Adrift

We drilled a great hole in the ocean
Midst waves and with endless commotion
But why are we doing it,
God knows, just keep screwing it,
To the basalt and beyond is the notion.

—Poem written by a crew member of the *Glomar
Challenger* in 1972, quoted in David J. Kinsman,
"The History of Leg 24"

Twenty-five hundred meters below the surface of the ocean, four hundred miles off the South American coast, the near-freezing temperatures of cold seawater give way to the shimmering, scalding waters of a hydrothermal vent, where a mineral-laden soup belches upward from the rock, superheated by magma below the surface of the floor. The hot fluids bloom upward and expand; sometimes the contents are black, and at other vents they are clear.[1] There is movement all along the chimney. Spiny and pinkish in color, shrimp swarm together as they roam over homes dappled with tube worms, mussels, and rugged rock. They are blind; unbothered by the darkness or the heat. Like all inhabitants of seafloor vents living in deep dwellings, heated hydrogen, sulfur, and methane-rich vent fluids are the primary source of energy. Many species of shrimp and gastropods live here in symbiotic existence with chemoautotrophic vent bacteria, collecting them from the sides of active smokers with their gill chambers and appendages and then ingesting them or taking up organic compounds through epidermal transfer.[2] Meanwhile, the bacteria get a prime location between vent fluids, which are rich electron donors, and the nearby seawater.[3] These critters live in a plume. And now, they may die by a plume, though it will not be the devil they know.

Hydrothermal vents are nestled in the bathypelagic zone of the ocean, where depths reach anywhere from 1,000 to 4,000 meters or more.[4] Along with crustaceans and mussels, these structures are home to polymetallic sulfide nodules, making them one of three types of benthic ecosystems containing mineral resources that are valuable to humans. In quieter waters like the abyssal planes of the Clarion-Clipperton Zone (CCZ), 4,000 to 5,500 meters below the surface of the Pacific, or on the slopes of underwater mountain ranges, the deep seafloor is studded with nodules and crusts containing manganese, cobalt, copper, lead, zinc, silver, rare earth elements, and gold deposits.[5] Commercial interest in seabed mining began decades before the capabilities for retrieving minerals existed, reflecting an imaginary of the deep sea as a supplementary space for modern terrestrial existence. In 1960, Leonard Engel wrote a piece for *Harper's Magazine* in which he muses, "Some day, both the nodules—whose origin is unknown but which contain manganese, copper, cobalt, and nickel—and the sea water will be mined. Oceanographic research can also determine whether the sea bottom is a safe resting place for atomic wastes."[6] Speculation like this about the emergence of a global deep-sea mining frenzy helped fuel the last three decades of international regulation, scientific research, and environmental activism around deep-ocean environments. This includes, notably, the 1982 United Nations Convention on the Law of the Sea (UNCLOS).

In the years since, demand for metals used in the production of digital technology and green energy technology has surged in the Asia Pacific, while automobile manufacturing in countries like the United States, Germany, and Japan drives the market for rare earth minerals and metals like copper, gold, platinum, and nickel. These resources are found in greater quantities at the ocean bottom than on land, where deposits are shrinking and increasingly contested. The so-called greener future is really a browner, muddier one: the minerals within seabed sedimentary deposits signal to us the next way civilization will build itself up. Not brick by brick as before, but windmill by windmill, battery by battery, Tesla by Tesla. Roopali Phadke calls this trade-off between environmentally destructive mining and environmentally destructive carbon emissions the "green energy bargain."[7] Seabed mining may very well help wean humanity off of damaging carbon emissions from fossil fuels. Yet there is still a widespread scientific consensus that sediment plumes from mining activities will take life away and damage deep-sea ecosystems on timescales far longer than what humanity can perceive.

Despite early interest in mining, delays and difficulties around technological development, funding, and investment means that seabed mining has only approached viability in the last decade.[8] Today, stakeholders are in the process of assessing seabed mining's potentially enormous social and environmental consequences.[9] But as with any mining industry, there is a steep price to be paid. For the many scientists and local communities concerned about the environmental effects of these extractions, anticipated underwater mining plumes, created in the wake of mining vessels and the dumping of wastewater, have become the primary objects of anxiety for understanding the impact of deep-sea mining on deep-sea ecosystems.[10]

Plume research and visualization, which enfolds both spectacular ("fast") forms of violence as well as less visible ("slow") ones, has had a marked impact on the development of the deep-sea mining code as well as applications of environmental justice in this context.[11] A critical media studies approach to thinking through plumes understands that scientific visualizations, models, and sensed data are non-neutral endeavors. Toxic plumes might alternately be depicted and thought of as invisible dangers, sources of risk for investors, visible symptoms of larger and more intangible events, or natural phenomena that can be controlled through dispersion. In this sense, plume science is permeated by cultural ideologies and by questions of political economy. Recruited into a broader media history of the Anthropocene, plume mediations have a direct impact on the fate of deep-sea biomes and, perhaps, on life as we know it. Recognizing these stakes, this chapter examines how the scientific study of underwater sediment plumes participates in the social construction of risk more broadly.

We often consider resource extraction from a terrestrial point of view. And perhaps there is an odd sisterhood between water and rock, between deep-sea mining and terrestrial mining, that precipitates from narratives around the seafloor. The Earth's crust is, after all, contiguous between the land and sea. Sediment plumes transport bits of land into the sea, throwing boundaries between settlement and sediment, endurance and erasure, into turbulent disarray. Yet, a simple comparison, or the use of terrestrial metaphors to describe the aquatic, can be inadequate, erasing much of the labor of extraction itself. The challenge, then, is to articulate a relationship between multiple sites of turbulence without losing sight of the ocean's unique material qualities.[12] Turbulence itself is a natural mediatory process, aiding the transduction of material from one state to another and in the transportation of signal and nutrients. From the late Latin *turbulentus*, meaning "full of disturbance or commotion,"[13] *turbulence* is a scientific term to describe the energetic, eddying motions found in everything from clouds to smoke to ocean waters. Such rotations are highly effective in dispersing material and transferring energy, heat, and solutes like oxygen or nitrogen.[14] Baked into this definition is thus a role for fluids in the movement of matter and energy.[15]

We can think once again in terms of elemental mediation, which allows us to consider processes of dispersion, dilution, containment, and other transformations within the conceptual framework of mediation. In science, the elements are thought of as chemical building blocks of life, and in a cultural context that sense of the elemental as elementary is maintained. What we think of as the elements—fluids, minerals, muds, winds, fires, and more—are themselves media that are productive "of feelings, of stories, and even, perversely, of life."[16] Pondering turbulent plumes as elemental media rather than as objects of representational media suggests a consideration of how basic substances like seawater, sediment, and slurry contribute to the strategies, devices, and environments through which humans make sense of and communicate information. It draws attention to the agency of seawater outside of scientific studies, and thus beyond the charade of containment that tends to characterize representational media about underwater environments. Like other media, turbulent plumes help to frame our understanding of the world and our relationships to it.

Analyzing scientific discourse alongside an elemental framework, I will argue that when it comes to sediment plumes, definitions of underwater hazard and risk are often camouflaged by ideas of resilience, as well as mediated by a politics of precaution. Echoing chapter 1, human understandings of risk, precaution, and hazard are key to the production of the seafloor as a resource frontier. To unpack these discourses, I will explore the science and mediation of plumes, including old and new techniques for modeling underwater plumes. By exploring studies of the Deepwater Horizon oil plume as well as attempts to model deep-sea mining sediment plumes, I will consider how extractive industries have influenced the scientific and technical knowledge that is produced around these perilous formations. The history of how turbulent plumes have been visualized and understood is full of friction and contestation, reflecting the different stakeholders involved. A distinct yet disjointed public is oriented around plumate media. Our knowledge of the plumes themselves is also determined through these same social networks. Finally, the figure of the plume is not just a physical phenomenon but also a potential model for thought. What can be found in a mining plume? Basaltic rock, seawater, detritus, minerals, toxic chemicals, heat, bits of animal carcasses, a means of making money, a chance to end the climate crisis, the defiling of an imagined wilderness, humanity's hopes for the future.

Plumate Mediations

Underwater sediment plumes are brought to the public eye primarily through photographs, diagrams, satellite images, and models. These mediations comprise a field of evidentiary knowledge that is tied to unique cultural modes of representation, through which plumes become environmentalist icons and signifiers. Plumes have long been used as visual indexes of trouble and destruction. To this day, we continue to associate expanding tendrils of fluid and smoke with events such as war, fire, and contamination. From the billowing black clouds above industrial smokestacks to subsea oil blowouts, buoyant fluids and vapors have a privileged place in human encounters with environmental pollution. Today, efforts to model, track, and understand the movements of plumes have expanded in their relevance; they are seen as potentially valuable for anticipating underwater avalanches and threats to human infrastructure, predicting the location of extractable sediment deposits, cleaning up pollution, as well as, of course, assessing environmental impact.

Although we think of them as visual objects, plumes are spectral, crossing borders between the perceptible and the imperceptible. They are defined through their perpetual motion, their propensity to travel, expand, and disappear. They are not only elemental figures, but transelemental, born out of the intermingling of two or more substances—

typically sediment, air, or water. This means that, like sound, plumes are disruptive, thwarting boundaries, capable of both threatening and cultivating life. Andrea Ballestero, writing on aquifer pollution, calls the plume a "guest that dissolves into its host, at different concentrations throughout its elongated figure."[17] For Ballestero, what is distinct about underwater plumes is their *indistinction* from their substrates. They cannot be separated out at a material level, but are made sense of through movements and concentrations. Where solutions imply stability and clarity, plumes, as Ballestero notes, imply change and shape-shifting.

The tension between the spectacle and spectrality of plumes generates distinct collective reactions to environmental risk. Elizabeth Povinelli writes that living among toxic plumes "sparks an affective fire we call *anxiety*, which sears the neural system according to its own logics and remedies, adding velocity to system." She suggests that as a result of this anxiety, "quasi-events and quasi-substances" like fires, fogs, and winds call for enhanced forms of perception or perceptual techniques.[18] Indeed, a toxic plume is something you can predict and sometimes see, but it cannot be fully grasped without technology. Representations of toxic plumes serve to "inoculate" us to their perceived environmental and health risks.[19] Sensors, dyes, and other devices attempt to maintain control over their mobility, making visible that which is always in the process of becoming invisible.

The project of making plumes visible implicates both a human spectator and an ideal scale of visibility. Elemental theorists like Jeffrey Cohen have argued that one of the affordances of thinking through elements is that they bring a "human-scale entry into nonhuman relations."[20] Similarly, media representations find human-scale entry points in the way they structure conventional ways of perceiving and understanding indistinct objects like plumes as graspable visual objects. This works alongside descriptive text. For instance, terms like "lobes," "bulges," and "fingers" are often used to describe and embody plume formations and settlement patterns.[21] Plumes also appear in a multitude of representational forms (models, illustrations, images) which have an inherent affinity for contours and containers, in addition to data sets and graphical media.[22] Taken together, descriptive and visual practices amount to a distinct plumate scopic regime, an ordering and shaping of the "productive, cognitive, and desiring capacities of the human subject."[23]

Scalar translations of plumes do not just operate spatially—they make changes over millennia tangible through estimates about resilience and recovery. Yet, for our purposes, there are affordances to returning to these deeper timescales for mediation. Circulating from the seafloor to the surface, materials like slurries, sludges, and sediments elude individuated identities, naming instead large-scale relations between human actors, aquatic life, rock, and technology. Enfolding such relations, Jussi Parikka's *A Geology of Media* usefully extends the concept of media beyond machines to their geological stories, while mobilizing a media materialism that lends ontological and epistemological agency to technologies outside of their human contexts.[24] Parikka's scholarship fixates on mobilizations and transmutations of geological matter from inert to vital contexts, as dead matter is enlivened and repurposed in mining. This expanded media ecology and "paleotechnics" is particularly befitting for the deep-sea context, where gold, copper, zinc, and rare earth minerals are used as raw materials for cell phones, electric vehicles, and other media devices that rely on lithium-ion batteries.[25]

This construction of a plumate scopic regime is key for the mining industry, which is charged with demonstrating its capacity to mitigate and/or contain the turbulence created by seabed mining. Analyzing promotional media produced by mining contractors, Stefanie Hessler writes that "renderings are designed to make prospecting endeavors seem both realistic and attractive for potential investors, while making the environmental impact seem negligible, as the cuts and incisions in the seabed are made to look clean and non-invasive."[26] The video that Hessler describes from Nautilus Minerals is a simulated rendering of the process, and it notably leaves out plumes, which muck up the illusion. More recently, contractors have been under greater pressure to account for sediment plumes, but their appearance is mitigated through an aesthetic of containment.

For example, Blue Nodules, a 2020 seabed mining prototype sponsored by the Dutch maritime technology company Royal IHC, produced an educational video about the mining experiment and process, including a photorealistic rendering of mining plumes (Figure 8). The video narration does not deny the environmental impacts of mining plumes, but minimizes them through editing. First, the simulated image does not show sediment plumes in their full expanse, with the narrator explaining

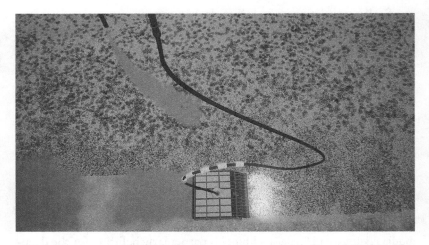

Figure 8. An image of the Blue Nodules mining prototype from their video summary on YouTube, posted in August 2020. https://www.youtube.com/watch?v=pCus0hTsibc.

simply that "the finest grains are transported over much greater distances." Meanwhile, passive grammatical formations, external comparisons, and formal language divert responsibility for ecological damage and marine death: "Similar to other extractive industries, small animals living within the sediment and attached to the nodules will be removed." The discussion of plume impacts is then bookended by detailed explanations and images of the mineral filtering and pumping processes, including enlarged renderings of nodules contained within the jumper hose: "We found that only a small fraction would be lost."[27] This effectively returns the viewer to a high-tech prospecting gaze, implying overarching values of efficiency and maximum profit. In the next section I will explore plumate visualizations from a historically situated perspective, complementing this industrial gaze with one that has been constituted by academic researchers.

Figuring the Plume

Research around turbulent motions in fluid dynamics has helped produce the plume as an abstractable figure for study. Figures, unlike symbols, are characterized by what Erich Auerbach describes as "change amid the enduring essence, the shades of meaning between copy and archetype."[28] This simultaneous impermanence of form and endurance of essence is what allows figures to act as contingent and mutable stand-ins for some aspect of a phenomenon across different kinds of mediation. While they may differ drastically in their makeup, science has enabled buoyant plumes—whether man-made or natural, aerosols or fluid—to become figures in this sense. Their behavior is replicable on micro-, meso-, and macroscales, and thus they can be easily studied under experimental settings and used to explain like phenomena.

Soviet theoretical physicist Yakov Borisovich Zeldovich was a pioneering agent in plume modeling, studying the motion of submerged jets. His 1937 paper "The Asymptotic Laws of Freely-Ascending Convective Flows" was published just a few years before his groundbreaking involvement in the development of the Soviet Union's nuclear bomb project in 1940. A decade and a half later, plume modeling grew substantially with the application of new mathematical techniques. In particular, the macroscopic (MTT) approach, commonly credited with launching the field of classical plume theory, was popularized as a "theoretical description" that "encapsulates the turbulent engulfment of ambient fluid into a plume."[29] MTT greatly simplified existing theoretical plume models by making three assumptions about the entrainment of plumes (entrainment refers to the proportional speeds at which the fluid draws along itself), allowing researchers to solve ordinary differential equations for any given plume source. Theoretical models such as these did not fixate on the materiality of the plume itself, whether gas, liquid, heat, or chemical. However, additional applied studies around the environmental impacts of toxic plumes were soon to follow.

In the United States the 1960s environmental movement brought with it a new tide of research on air pollution, including studies of buoyant plumes of waste and heat from smokestacks, chimneys, and automobiles.[30] There was a strong social impetus for this research, as it took place shortly after the publication of Rachel Carson's influential *Silent Spring* in 1962, the implementation of seminal environmental regulations (including the 1963 Clean Air Act, 1965 Motor Vehicle Air Pollution Control Act, the 1965 Water Quality Act and the National Environmental Policy Act of 1968), and a "Whole Earth" ideology that, alongside satellite and space photography, emphasized planetary fragility and interdependence.[31] While much was made of gaseous plumes in terrestrial environments during this time, the ocean was a less visible yet significant site for the development of plume theory and experimentation with plume tracking. For instance, after the 1977 discovery of hydrothermal vents by the crew of the Galápagos Hydrothermal Expedition, many early fluid dynamical modeling efforts used buoyant plume flows from hydrothermal vents as their basis.[32] The insights from these observations were integral both to fluid dynamics and to oceanography as a whole. Scientists considered the role of these natural plumes in transporting chemical signals, nutrients, and biological material, which fed into the systems-oriented thinking that characterizes the earth sciences today.[33]

The practical study of plumes has often toggled between examining natural plume formations like those found at hydrothermal vents and anthropogenic contaminations. However, social anxieties over environmental toxicity have often taken the driver's seat when it comes to pushing this research forward. One event in particular, the 2010 Deepwater Horizon oil spill, is notable for advancing the techniques in tracking plumes through remote aquatic environments. Shortly after the spill, a group of marine biogeochemists from the Woods Hole Oceanographic Institution embarked on a mission to document the spreading of a subsea plume from the oil spill, making it one of the first attempts to hunt for oceanic plumes of this kind.[34] There was a material challenge to this project, in that the Deepwater Horizon hydrocarbon mass was clear, odorless, and difficult to detect. In the past, plumate scopic regimes were more forensic and focused on studying the traces left from industrial crimes like oil spills. This typically meant tracking indirect effects, such as searching for dead zones, biological changes, and oxygen depletion. Tracking a live plume, however, was like "trying to map a smoke trail in a vast sky somewhere below the clouds by dangling lines with smoke detectors from a helicopter far above the clouds."[35] This difficulty thus drove the production of technologies that focused on visibility and control, from the use of tracking dyes, to autonomous sensors, to modeling on computers.

To study the Deepwater Horizon plume, researchers developed a trackable autonomous underwater vehicle called *Sentry*, which was then fitted with sensors and a portable mass spectrometer called TETHYS.

The combination of technologies created what the researchers referred to as a "powerful oceanographic bloodhound."[36] This platform was towed by a ship horizontally along a set path, while bobbing up and down to cover breadth and depth, tagging data points at individual locations and transmitting that data back to the ship in real time. The findings showed that the plume existed, and it also gave oceanographers information about how quickly the plume moved from the blowout, how far, how persistent it was before biodegrading, and how plumes move differently in deep as opposed to shallow environments. Meanwhile, the language of the hunt, including the reference to an "oceanographic bloodhound," transformed the plume itself into something like prey or a criminal body. As plumes become lively actors with a particular behavior that must be captured or brought to heel, oceanographers become ocean detectives and police. The emergence of plume modeling has been driven by an additional desire to improve cleanup and recovery of toxic plumes by anticipating and profiling their movements. Beyond simply fishing for plumes, oceanographers started to conduct experiments, creating and observing artificially generated plumes with tracer dyes, or modeling plumes by creating contained underwater avalanches in aquariums.[37] These smaller plumes might be understood as figures and stand-ins for more consequential plumes such as oil spills.

Due to the multiscalar challenges of how toxic metals and chemicals move through water, mediations of oil plumes are closely connected to technologies of containment. Infamously, the Deepwater Horizon oil spill was treated with a chemical dispersant called Corexit, which was administered through aerial spraying. Corexit works to emulsify oil into droplets, which would in theory allow the oil plume to biodegrade faster. Here, containment of the catastrophe was funneled into an act of dispersion. Melody Jue uses Corexit as an exemplary case study for what she describes as the "fluid cut," expanding on film theories of the cut to account for the antivisual logics of dispersants like Corexit and Dawn dish soap. Jue seizes upon the ways in which surfactants have been involved in obfuscating the visual residues of oil and, by extension, British Petroleum's social responsibility. She explains: "By describing the material aesthetic of the fluid cut, I show how Anthropocene discourse should not only involve geologic accumulation but also chemical agency, addressing processes specific liquid and their capacity of cutting, dissolving,

dispersing, and suspending."[38] Jue's turn toward antivisuality here resonates with my proposal of dark mediation, which describes the limiting
of signals (visual or otherwise) through media technology and a decoupling of mediation from the values of transparency. In her analysis, Jue
uses the term "negative mediation" instead to describe this informatic
erasure. While negative mediation shares dark mediation's focus on the
politics of obfuscation, it also goes a step further to think specifically
about elemental media like glaciers, tree rings, and migration routes that
are destroyed in the wake of the Anthropocene.

Efforts to fight the plume from within the plume with Corexit
fell short, and later it came to exemplify technological solutionism gone
wrong, as retrospective studies of the BP oil spill cleanup workers found
that Corexit was acutely toxic to people, leading to "significantly altered
blood profiles, liver enzymes, and somatic systems."[39] Other studies
also found that the dispersant was highly toxic to marine organisms like
microzooplankton—far more so than crude oil alone, resulting in significant disruption to natural food chains in the Gulf.[40] Contrary to what
proponents had suggested, the chemical actually made biodegradation
more difficult. What this case demonstrated was the importance of taking action to prevent spills, rather than merely addressing the aftermath
of such man-made ecological disasters. While WHOI's plume-tracking
study contributed important information on how aquatic plumes move
and biodegrade, one could critique it as yet another "solution" that did
little to prevent or remedy the disaster itself.

Tracking Sediment Plumes from Seabed Mining

Beyond oil spills, seabed mining is now providing a compelling practical
application for the tracking and modeling of underwater plumes. Mechanisms for deep-sea mining will produce two types of sediment plumes:
collector plumes, which are created through the physical movement of
the collector machine on the seafloor; and wastewater plumes, referred
to as "dewatering plumes," which are returned to the sea after the valued
polymetallic nodules are vacuumed up. In addition to this, each plume
can be characterized in terms of a dynamic plume (the initial disturbance)
and an ambient plume that comes after.[41]

Both collector plumes and wastewater plumes have been the subject
of in situ experimentation. During the initial wave of innovation around

commercial seabed mining, scientists conducted controlled-impact experiments that attempted to replicate anthropogenic disturbances and then observe the movements of sediment plumes. Many of these piggy-backed onto proof-of-concept experiments for deep-sea mining itself, including the 1976 Deep Ocean Mining Environmental Study (DOMES), which used the research vessel *Oceanographer* to test techniques for poly-metallic nodule mining and study the subsequent resuspension of sediments.[42] DOMES tracked of the size and dispersion of the two kinds of mining plumes, while also collecting biological data before and after the disturbance.[43] To study the plumes, the DOMES group used drogues, or trailed sampling devices. These were simpler than but similar in principle to the approach used for the Deepwater Horizon spill. Specifically, dewatering and bottom sediment plumes were tracked with a series of traverses by a nephelometer, an instrument that measures the concentration of suspended particulates by their light-scattering properties, over the course of its existence. Similar projects occurred throughout the early 1990s in the Indian Ocean, off the coast of Japan, the Peru Basin, and the CCZ.[44]

Roughly a decade after DOMES, Scripps researcher Gary Taghon submitted another proposal to conduct a controlled-impact experiment to track sediment using tracers rather than towed sensing equipment. Specifically, the experiment sought to label sediment with inert gold tri-chloride, activate it with a nuclear reactor, and then quantify that radio-active gold.[45] The goal was "to provide the means of tracing sediments that have been taken from the deep seabed and redeposited in some manner to simulate the benthic impacts that would be associated with the deep seabed mining of manganese nodules."[46] Like DOMES, this proposal sought to study the environmental impacts of deep-sea mining by producing anthropogenic disturbances that re-create those environmental impacts.

Recent studies of turbulent plumes build from this tracer method. Most notably, a 2017 MIT study called PLUMEX was conducted in conjunction with the International Seabed Authority (ISA) by a team of researchers including MIT oceanographer Thomas Peacock, Scripps oceanographer Matthew Alford, fluid dynamics specialist Carlos Muñoz Royo, earth scientist Glenn Flierl, and mechanical engineer Pierre Lermuseiaux. Using the research vessel *Sally Ride*, PLUMEX generated, monitored, and gathered in situ data about a turbulent plume of water that would resemble the dewatering (waste) plumes created during the mining process. This experimental plume consisted of ninety thousand gallons of sediment-laden water and actual deep-sea sediments, marked with a bright pink tracer dye. This sediment water was then pumped twenty-five to one hundred meters below the surface.

In a video introducing the project, Peacock explains, "The goal is to understand if a sediment plume is released from a surface mining vessel, how far and in what concentration will that sediment travel? We can use that information to then determine how that will impact the ocean biology."[47] The main components of the project (see Figure 9) included the research vessel, the phased-array Doppler sonar (PADS), an ROV fitted with a CTD (conductivity, temperature, and depth) monitor and fluorometer to monitor characteristics of the plume, and the proposed depth at which the experimental sediment plume will be released. Royo reflected on this work in an interview and in particular on the need for

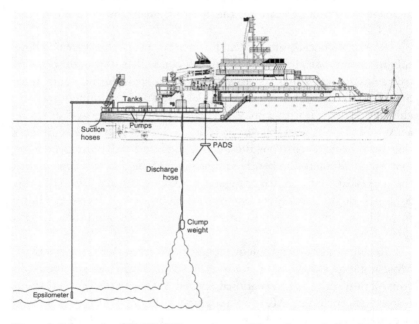

Figure 9. Schematic of PLUMEX experiment. The sediment-laden plume is depicted with monitoring from PADs acoustic technology mounted to the research vessel *Sally Ride*. Figure by Scripps Institution of Oceanography. Image copyright and courtesy of the Massachusetts Institute of Technology (MIT).

interdisciplinary collaboration between fluid dynamics, oceanography, and biology as a precondition to understanding the metrics and objectives of plume research. He also spoke about the challenges of having to work at a faster pace than traditional research due to the timelines set by mining corporations. Although deep-sea mining contractors provide scientists with most of the data about the deep sea, few take the effort to measure turbulence, and thus turbulence sensing requires partnership with academic oceanographic research institutions like Scripps.[48] However, Royo and his colleagues remain acutely aware of the economic and geopolitical stakes of their research. Researchers interviewed in the video summary of the project attempt to articulate its practical importance and are shown in conjunction with historical and contemporary footage of international legislative meetings.

The video starts with Peacock's statements: "I think the question of deep sea mining is part of a bigger question of how the world is going to sustain itself in the next ten, twenty, thirty, forty years. We want electric powered vehicles, we want solar cells, clean energy, but all this requires resources. . . . The question is, can we do a better job in the ocean than we do on land?" Immediately, what stands out about this remark is the short-term timescale for sustainability. Forty years, after all, is a drop in the bucket when it comes to the age of the seabed and the environmental effects that are certain to accompany mining. In fact, in a 2018 *Scientific American* article, Peacock and Alford write, "Given that nodules take millions of years to form and that biological communities away from hydrothermal vents in the deep ocean are very slow to develop, harvested regions are unlikely to recover on any human timescale."[49] As such, Peacock's choice to define sustainability in terms of mere decades in this video is a telling reminder that when it comes to seabed mining, short-term gains for humanity continue to hold the most sway in the public imaginary.

Peacock and other marine scientists are often positioned as translators between stakeholders. And in many cases, ISA affiliates like the PLUMEX researchers express a certain resigned ambivalence about seabed mining. This is a particular kind of resignation that characterizes mainstream mediations of the Anthropocene. Environmental scientists tend to view markets and their mandates for extraction as inevitable. With the present-day normalization of finance capitalism, even imaginaries of a sustainable future are built on the premise of environmental destruction.

While most maintain enough critical distance to understand that these extractions will result in damage to the planet, they continue to invest in the ideology of the blue economy.[50] The corporate discourses and mediations around seabed mining that I will discuss later in this chapter, from resilience to risk management, fundamentally embed this remorseful resignation toward extraction and its accompanying foreclosures of the future.

As mining has evolved, deep-sea researchers have become increasingly aware of the importance of articulating their findings to the ISA and embedding them into environmental guidelines and regulations. Cindy Dover, one of the first ISA-affiliated deep-sea scientists working to establish environmental baselines for seabed mining, explains that her sentiment at the time she was offered the consulting role was that "it's not going to happen for a long time," noting that initially she did not take the plans to mine seriously. Times, it seems, have changed: "There were a lot of negative views of scientists working with a mining company and I had gone to the dark side. Now, everyone is working with mining, every paper I read says that we need to do this because we need to understand the impact of mining."[51] Indeed, while organizations ranging from NOAA to the EU have funded seabed projects, the steep financial costs of deep-sea research means that the vast majority of this exploratory work is done by industry contractors.

Couched within this research ecosystem, Royo explained that PLUMEX ultimately was an important project that addressed the midwater plume—an area in the water column that was previously understudied. However, there have been fewer studies of the near-bottom sediment plume, which is linked to the action of the collector machines at the bottom. Information about the benthic plume is mostly known from studies that occurred alongside mining experiments in the 1970s and 1980s, before the direct tracking and sampling of benthic plumes was possible. These studies examined the resettling of sediments after the collector's movements, noting higher particulate concentrations at depth than at the surface—a thicker plume.[52] At the seabed, the collector vehicle creates much more turbulence and thus a wake, which will create a turbidity current that propagates under its own weight. At the time of interviewing, Royo and his team were in the process of developing plume models for this deep seabed zone.[53]

There are notable similarities between the political commitments of research about deep-sea mining plumes, and the research on oil spill plumes that came prior to this. For both, the scale of the plume's movement is the primary question and source of anxiety, while dispersion and resilience are the metrics that assuage fears and provide a route of reconciliation between industry, policymakers, and the wider plume public. In the mining case, the observed distances that plumes travel will influence the regulation around how far mining activities must take place from each other as well as at what intervals they can occur to allow time for ecosystem recovery. Plumes dispersed at the bottom are thought of as more environmentally friendly than dispersion at the water column, due in part to high pressures, slower currents, less spread, and diminished human reliance on deep-sea ecosystems.

While environmental activists might find it easy to critique scientists who perceive deep-sea mining as an inevitability, this work is driven by an important nugget of hope. Past efforts to model plumes tended to aid in mitigation efforts, but they were not necessarily preventive and often did not question the extractive sources of toxic plumes. Unlike studies of natural plumes or oil spills, however, the stakes of mediating underwater plumes change dramatically with deep-sea mining, as experiments around sediment and waste plumes are coming before mining itself. This means that information gathered about plumes now may inform regulations and limitations on mining later. It is during this unique window prior to the formation of the mining code and commencement of large-scale commercial seabed mining that discourses of risk and precaution come to the fore, supplanting a forensic or reactionary perspective on environmental pollution and toxicity. Echoing legal adviser and international arbitrator Christopher W. Pinto, ocean governance expert Alina Jaeckel observes that "the ISA has been described as 'the result of an uncharacteristic attempt by the States to adopt a precautionary approach and ensure rational and equitable utilization of the resources of the seabed through regulating the mining industry before the need for regulation had been generally recognized.'"[54] At present, scientists carry most of the burden in carving out this precautionary path, which, most importantly, will include the designation of Marine Protected Areas as well as regulations on the frequency of mining activities in any given area.

Plumes as Precautionary Mediations

While benthic researchers often use the terms "precaution" and "risk" interchangeably, there is an appreciable difference between the two ideas. Media scholars and sociologists have used the term "risk society" (coined by Ulrich Beck) to describe modern society as increasingly preoccupied with the transformation of uncertainty into predictable, routinized formats. In the context of a digital mediasphere, critical media scholars have taken this point further to describe "risk media" as a distinct genre of media which includes predictive tools like computer modeling, probability calculations, and algorithms. Risk media, including the nascent technologies developed to model toxic plumes, may communicate perceptions of risk while also replicating or intensifying the affective experience of risk.[55] Perhaps most significantly, these tools also allow risk to be monetized, and increasingly, risk is understood through its attachments to reward.

In her critique of finance capitalism, Aimee Bahng considers the limits of corporate risk mediation and its aspirations to tame uncertainty: "Uncertainty cuts loose from risk discourse's capture, eluding containment and quantification. While true uncertainty might refuse the grid of intelligibility that securitization would foist upon it, it remains knowable as lived experience, felt and negotiated perhaps most profoundly by those held in 'the waiting-room of history.'"[56] Bahng articulates a relationship to risk from the perspective of the subaltern— from those who are the casualties of corporate profit seekers—offering a queer, migrant futurity that might transform the "waiting-room" into a horizon. In short, she aims to broaden the prospect by returning to how uncertainty is felt and reacted to from the margins and from below. In this vein, when it comes to contestations over deep-sea mining, precaution emerges as both a practical tool of resistance and a meaningful reorientation toward uncertainty and futurity from the bottom.

Although risk and precaution are both ultimately speculative ways of thinking about the future, precaution is a response that cannot be reduced to the control of uncertainty through scientific enumeration or statistics. Molly Wallace explains that instead, precaution emphasizes "uncertainness, unpredictability, and incalculability, and a precautionary reading practice might offer a mode for risk critics, interested both in tracking the ethico-political consequences of risk and potentially forestalling catastrophe."[57] That is to say, in the absence of knowledge, precaution ties speculation to an ethics of accountability. In the deep sea, where humanity's reach is sparse, this absence of knowledge is essentially a guarantee, as these areas are largely unexplored. Unsurprisingly, scientific development of a plumate scopic regime for deep-sea mining is most frequently digested as a form of risk analysis. But as plumes are both scientific and social figures, sediment plume research is also linked to a sense of social responsibility and to precaution. This is most explicitly manifest in the precautionary principle, which has been a key regulatory tenet for ocean exploitation.

The precautionary principle, proposed at the 1982 UN World Charter for Nature, currently informs the ISA regulatory framework for mineral exploitation of the deep sea. It reads:

> All persons engaged in activities in the Area shall apply the Precautionary Approach, as reflected in Principle 15 of the Rio Declaration, to the assessment and management of risk of harm to the Marine Environment from Exploitation Activities in the Area and where scientific evidence concerning the scope and potential negative impact of the activity in question is insufficient but where there are plausible indications of potential risks of Serious Harm to the Marine Environment.[58]

This codification of precaution has played a part in positioning scientists as the interlocutors between corporate entities, states, and planetary ecosystems. Their expertise is used to seek out new extractive sites, guide, and justify policy for environmental impact assessment and enforcement. Given the manifold uncertainties around seabed mining and the deep sea, the precautionary principle is also frequently invoked by community activists to cast doubt over the economic case for seabed mining relative to its deleterious environmental and social effects.

Despite a cautiously optimistic and regulated start to seabed mining, the clock has run out for deep-sea researchers and communities to make their voices heard. Article 15 of UNCLOS was triggered in June 2021, starting a two-year countdown toward the completion of a mining code. In other words, Article 15 put a limit on the amount of time that deep-sea research could continue unhindered before global exploitation begins in earnest, *regardless of the code's completion*.[59] In this interim, researchers and policymakers were directed to gather as much information about the deep sea as possible and incorporate it into regulatory frameworks before the potentially catastrophic destruction of what many describe as the last great wilderness on Earth. Although the idea of wilderness is itself an unstable signifier in the era of the Anthropocene, when it comes to the deep sea it continues to exist as a powerful imaginary of environmental recalcitrance from human influence. However, as with existing terrestrial and offshore extractive industries, it seems likely that corporate and political interests in the short-term capital gains of mining will soon overtake attention to long-term views on planetary health and environmental destruction. The Article 15 deadline came and went without a finalized mining code, and while negotiations for nonbinding provisional agreements continue, so have plans for mineral exploitation.[60]

The reaction among deep-ocean researchers to Article 15's invocation was overwhelmingly one of concern. And even now, the precautionary principle remains the primary legal bulwark against a runaway deep-sea gold rush. Following the announcement, the Deep-Ocean Stewardship Initiative, registered as an observer for the ISA, issued this statement:

> Rushing the regulations to meet a two-year goal would run counter to the precautionary approach, which requires Member States to err on the side of caution. For example, the ISA has yet to agree on the overarching Strategic Environmental Goals and Objectives, define "serious harm" and associated adverse change, as well as specific criteria to operationalize, measure and monitor, and put in place effective regional environmental management plans.[61]

Benthic oceanographer Lisa Levin echoes these concerns, stating, "Even at its most intensive pace, science cannot yield the information needed to manage this nascent industry in only 2 years."[62] Time, it seems, has become the limited resource.

Given the looming deadline, research into sediment plumes produced through mining and their effect on the water column and benthic ecosystems is even more urgent. The modeling of plumes in the water column is accompanied by research on deep-sea ecosystems and their recovery timescales. This work focuses on what we can afford to lose before recovery is not possible, helping to answer questions like, how far can we go, or how much can we destroy? Here, a geographical question is mediated by the temporal scales of mining, which is itself a disputed terrain between precaution (a question of human responsibility) and resilience (nonhuman capacity for unaided recovery).

Take, for instance, the case of the Solwara 1 proposal by Nautilus Minerals, a Canadian company and the first to try to explore and mine hydrothermal vents commercially. The majority of seabed mining plume experiments are now done in the CCZ, an area of the Pacific just under the size of the United States where mining is most immediately viable. However, many early experiments, including environmental impact assessments for Solwara 1, focused on seafloor massive sulfide mining, which takes place at hydrothermal vents that are home to densely populated ecosystems—oases within the vast expanse of the seafloor. Initial environmental impact research at these vents led to recommendations around how to minimize plume generation, acknowledging a responsibility to refine the mining process itself. This includes recommendations to discharge the dewatering waste plume horizontally along the seafloor, closer to its origins, rather than at the water column, to minimize its spread.[63] These scientific recommendations put the onus of "precaution" on the technological minimization of plumes, advocating for delay and mitigation when technological solutions were absent.

By contrast, the public backlash to Nautilus Minerals and subsequent use of the precautionary principle was far more condemning of the industry as a whole. Nautilus received its mining permits in 2009, but over the course of a decade it found itself in a number of disputes with the Papua New Guinea government and with locals, drawing media attention and community ire. The Deep Sea Mining Campaign (DSMC), which represents community activists in Papua New Guinea, is one of many organizations that worked to halt the Solwara 1 project. Like other local groups, DSMC made the argument that mining contractors failed to adequately address social and environmental impacts and denied coastal communities a proper seat at the negotiation table: "This definition of the PP prioritizes the protection and well-being of communities and the environment. . . . Implicit in this principle is the social responsibility to protect the public from exposure to harm, when scientific investigation has found a plausible risk."[64] What is important about this statement is its emphasis on social responsibility as endemic to the concept of precaution, thereby extending it beyond the question of technological mitigation.

Invocations of the precautionary principle have continued to be used by anti-mining activists like Teina Mackenzie, a nonprofit leader involved with the Pacific Islands Forum. Mackenzie argues that mining has proceeded with a "single bearing, even if it goes by various names: progress, development, diversification, GDP, growth."[65] She references a legal opinion by C. J. Iorns Magallanes on the Seabed Minerals Act to advocate for a moratorium on mining, which contends that the decision to mine the seafloor was made with a dearth of data about the deep sea. While scientists can and do articulate risks to benthic ecosystems, public input is also important to definitions of "serious harm."

Despite the protections of the precautionary principle, international frameworks like those at the ISA also have the side effect of cultivating predatory relationships between Western nations, mining corporations, and island nations like Tonga or Papua New Guinea. Oftentimes, less powerful nations are pressured to legally sponsor and thereby provide legitimacy to foreign mining contractors from wealthier nations like Canada, even amid sustained opposition by members of coastal communities themselves. The Nautilus Minerals case is one of the few instances in which local opposition ultimately succeded in halting mining. But the greater outcomes of Papua New Guinea's partnership with Nautilus are much more bleak. Although Nautilus eventually went bankrupt in 2019, in 2011 its founders created another mining company called DeepGreen Metals (renamed The Metals Company in 2021), which now holds three exploration licenses in the CCZ. And while many early Nautilus investors became millionaires over the course of this debacle, Papua New Guinea's partnership with Nautilus left the country $125 million in debt.[66]

While there are certainly many intersections between scientists and environmentalists working with the ISA and the larger community of anti-mining protestors, the stakes and challenges vary dramatically across different groups. Katherine Sammler remarks that experimental mining and economic plunder "are of dire concern to many Pacific peoples, who have long experienced the effects of extractive practices of ocean and island natural resources."[67] For these communities, sediment and wastewater plumes will affect fisheries and generational livelihoods at sea, as aquatic pollutions cross borders. In her assessment of seabed regulation, for instance, Sammler points out the ways in which legal boundaries such as exclusive economic zones (EEZs) are "created in tension" with the ocean's materiality, which "spills over" political and legal boundaries.[68] In addition to the uncontainability of plumes, Rachel Reeves, in her study of the Seabed Minerals Authority (Cook Islands) public meetings, observed concerns about the legal enforcement of mining regulations and a historical lack of accountability around environmental mismanagement.[69] Even with the precautionary principle in place, it is clear that Pacific Islanders are poised to absorb enormous risks and damages from deep-sea mining.

Reflecting long histories of environmental injustice, activists in coastal communities sometimes invoke the plumes from oil spills to articulate mining hazards and challenge industrial contractors. For instance, in its efforts to protect Papua New Guinea waters from seabed mining, the Environmental Defenders Office, a community legal service based in the Australia Pacific, stated, "Just this year in the Pacific, we've seen oil and ore spills in Solomon Islands and the spillage of an estimated 200,000 litres of toxic red slurry from the Ramu Nickel mine in Madang, PNG."[70] More than individual events or objects, toxic plumes are used as symbols of the destructive capacities of extractive capitalism at large. They point to the deleterious impact of multiple kinds of terrestrial and aquatic extractions on both planetary ecosystems and local communities. From a social and political standpoint, precaution is not just a temporary brake—it is a state of permanent awareness that certain actions will have vast and unknowable consequences. As both a situated and expandable orientation toward the future, it also enables people to disrupt established pipelines of information and value between science and resource extraction.

The Politics of Resilience

The scientific negotiation of risks around toxic plumes has focused keenly on the scale of environmental and social impacts. Beyond studying the physical properties and movement of turbulent plumes, however, the other side of environmental impact assessment for seabed mining involves collaboration with deep-sea biologists to determine ecosystem resilience to mining plumes. Most deep-sea biologists agree that the deep ocean is a highly biodiverse, unique ecosystem that will be catastrophically affected by turbulent sediment plumes. I interviewed one of these deep-sea researchers, Craig Smith, who runs the benthic ecology lab at University of Hawaiʻi Mānoa, about the impact of mining in the CCZ. He explained that the concentration of suspended sediments and sediment accumulation rates are among the lowest on the planet. These are the clearest waters in the world because of the increased density at depth leading to a more stable state.[71] That means that turbidity events that might be considered mild in another context would likely be catastrophic to the sensitive filter feeders in the water column and at the bottom. Put bluntly, deep-sea communities are the least "resilient" ecosystems on the planet. And beyond that, biologists must take into account both acute exposure from the initial plumes, and chronic exposure. As Smith put it, "A lot of the impacts will take years to manifest themselves, and the recovery timescales will take even longer."

How then, does a marine biologist reconcile support for or work with the deep-sea mining industry, knowing its ecological effects? Sometimes, scientists choose to describe hazard in terms of distinct tipping points of change. Dover includes this idea in her own ethical negotiations around mining:

> I have a colleague that says if one nematode goes extinct, we shouldn't mine. I don't agree with that. But there's going to be some point where it makes a difference. . . . It's clear to me that you could remove a square meter of the Clarion-Clipperton Zone and you would not have a long-term effect on an ecosystem scale. What I don't know is how big you can destroy or degrade. How much can you do this stuff before we see something we care about change, like the oxygen concentration or the total biodiversity?[72]

The debate Dover describes is again a question of resilience, defined broadly as "a system's capacity to spontaneously reorganise itself in response to disturbance and adapt in ways that preserve its identity and function."[73] Here, resilience is used as a threshold of acceptability, beyond which lies a point of no return wherein the identity of this ecosystem is irrevocably damaged.

There are several important critiques of resilience thinking within the humanities. Melinda Cooper and Jeremy Walker's article on "Genealogies of Resilience" identifies the way the science of complex adaptive systems has led to a problematic merging of the ecological concept of resilience with neoliberal doctrines of governance. The underlying idea of an ecosystem's capacity to remain cohesive under perturbations has, the authors argue, bled across security, environmental, and infrastructural contexts, which have all moved away from the charge to prevent or avoid catastrophic events and toward the capacity to adapt. "Relying as it does on the non-equilibrium dynamics of complex systems theory, what the resilience perspective demands is not so much progressive adaptation to a continually reinvented norm as permanent adaptability to extremes of turbulence. . . . Resilience risks becoming the measure of one's fitness to survive in the turbulent order of things."[74] Cooper and Walker also point out that resilience has become a justification for neoliberal think tanks seeking to remove environmental protections, as the charge to create resilience appears to justify the assumption that adaptation thrives best without intervention. This puts the onus on the evolutionary idea of survival of the fittest, rather than on human responsibility to nonhuman others.[75]

Resilience rhetoric has already been weaponized with deep-sea mining as a means of camouflaging risk. Early industrial environmental impact statements often assumed the resilience of seabed communities because of the extreme heat and pressure of the deep sea, treating ecosystems such as hydrothermal vents like petri dishes that can be erased and reproduced. For instance, the Nautilus Minerals' environmental impact statement for the Solwara 1 project reads: "Recovery would occur once clean, hard surfaces emerge and new settlement occurs, as would be the case after a volcanic ash dump."[76] The comparison to terrestrial volcanic life also leads to assumptions about the natural hardiness of organisms, as Nautilus infers that "a high tolerance to metal concentrations on water and sediments would be of selective and survival value."[77] Solwara 1 mitigation proposals relied heavily on these insinuations about bottom dwellers, whose tolerance of high heat and metal concentrations are extrapolated to an idea of permanent adaptive ability. And this is far from the first instance in which obscure bottom dwellers have been used to mollify fears about extraction. Scientists have also frequently promoted microbes and bacteria as capable of cleaning up and dispersing waste like oil, metal, and plastic—an "all-natural" successor to Corexit that embodies biotechnological solutionism.[78] Unlike the charismatic whales that I discuss in chapters 2 and 4, crustaceans, microbes, and tube worms have historically enjoyed far less human empathy, and are thus arguably more vulnerable to commodification in these terms.[79]

More importantly, these assumptions about resilience muddy the physical character of the deep seabed. Solwara 1's equation between terrestrial and deep-sea environments has been roundly criticized by independent scientific researchers. The DSMC wrote: "By comparing apples to oranges, it is hardly surprising (but meaningless) that Solwara 1 is rated by EE as having a lower impact than the selected land-based mines on terrestrial values such as ground and fresh water quality, air quality, pollination, soil formation and retention, and recreational activities such as hiking and bike riding, and loss of agricultural land."[80] In addition, vent organisms clearly do not tolerate all kinds of extremes, and the very idea of extremity is relative. In one 2016 study, Auguste et al. demonstrated that shrimp (*R. exoculata*) adapted to live in a metal-rich vent-field environment were still highly sensitive to copper exposure in solution. Likewise, vent mussels from the "Lucky Strike" hydrothermal vent site that were exposed to copper solution experienced "elevated lipid peroxidation at copper concentrations in excess of 300 µg l^{-1}, indicating lipid membrane damage within these tissues."[81] The ability to live in an extreme environment in one sense thus does not presume a lack of vulnerability to all other kinds of temperature and chemical extremes—especially when it comes to anthropogenic disturbance.

Fundamentally, Solwara 1's impact statement also ignores and makes unfounded assumptions about the temporalities of resilience and anthropogenic disturbance, which, despite its frequently rapid physical manifestations, is also a form of slow violence.[82] A letter to the editor of *Nature Geoscience* written by Dover and twelve other researchers, oceanographers, and lawyers warns that given glacial rates of recovery, the existing regulatory frameworks for deep-sea mining industries prescribing solutions to biodiversity loss through avoidance and minimization (for instance, patchwork extraction to reduce the mining footprint), remediation (the halting or reversing of environmental damage), and offsetting (like-for-like counterbalancing of environmental impact) constitute "unrealistic" and "unattainable" goals.[83] Can a centuries-long scale of recovery described as "nearly forever" by scientists really be called resilience, or does it erode the very utility of the word? Is there a temporal limit that marks the boundary between turbulence and catastrophe?

A fluid environment like the ocean floor will always shift over time and experience periodic upheaval. What resilience arguments mask is the massive production of turbulence in another form, one that operates on a larger scale of disturbance than the perturbation of a single ecosystem. Hessler emphasizes this in her own discussions of ocean prospecting, arguing that "mining is justified by conceptual shortcomings that assume the seabed would stay the same and that the environmental effects can be localized."[84] Indeed, the Solwara 1 environmental impact statement willfully confuses different scalar impacts. If catastrophe signals a radical, irreversible loss of life or identity,[85] resilience grows in the shadow of certain yet "non-catastrophic" disaster. This orients the conversation away from preventing acute turbulences from radiating through deep temporalities and into other lands, and from the activities that add up to catastrophes like extinction or irreversible sea-level rise. Plumes exist on all scales—one cannot simply excise a single scalar perspective from the others. Beyond acknowledging the dynamism of environments, scalar mediation is key to grasping how technological interventions create radical environmental changes through time.

Cross-scalar mediations are endemic to environmental impact research itself. Since the late nineteenth century, the earth sciences have developed several methodologies for contending with multiple scales of environmental activity and impact. This includes methods of scale analysis—understanding that different phenomena occur at different scales and require particular sets of instrumentation—as well as downscaling, the alternation between global and local perspectives.[86] Oceanographers in particular use space-time diagrams to consider overlaps between biological and physical processes and the coupling of ocean processes over multiple scales. Today, such scalar mediations are central to the production of knowledge about climate change, which invite us to "imagine unconventional forms of causality and collectivity."[87]

While a sense of both place and planet has become imperative to forming international policies that might mitigate planet-wide environmental crises, the making of scalable environmental knowledge is, at the same time, intimately wrapped up in the history of Western imperialism and empire building.[88] The same can be said for plume research as well, which did not emerge out of nowhere. From smokestacks to oil spills, these scientific studies are outcomes of both environmentalist efforts and the violence of extraction, which has always been part of the price tag of settler colonialism. Just as climatology once helped European colonizers make decisions about where to pursue settlements in distant lands, plume research is helping neocolonial players decide where and how to place mines, cables, offshore drilling rigs, and other infrastructures.

What is being negotiated with mining plumes is ultimately the temporal and spatial boundaries of what constitutes hazard in the first place, which is bound up in the question of what is "natural" and what is "unnatural" with regard to turbulence (with "natural" having a de facto positive value). In the Nautilus case, resilience thinking diminished the very concept of turbulence as it positioned it as a norm and as natural, shifting the context from industrial pollution to antiquated notions of chaotic wilderness. Rather than solving the root causes of turbulence, the problem became about optimization and adaptation. This is concerning, because optimization orients the conversation toward how to live with and even benefit from uncertainty and turbulence. As Orit Halpern puts it, "It is not about a future that is better, but rather about an ecology that can absorb shocks while maintaining its functionality and organization."[89] When used to speculate about deep-sea mining, resilience thus assumes a path that is already in motion. The reality, of course, is that there is no magic threshold of hazard that will be universally agreed upon as acceptable, and that not everything "natural" is necessary or beneficial to the oceans as a whole. And so, we will have to ask existential questions about long-term human relationships to the environment, as well as ethical questions about the rights of nonhumans to exist outside the realm of our use and our perceptibility.

One substitute for resilience might be Nassim Nicholas Taleb's "antifragility," a measurable metric that puts focus on the protection and sustenance of futures as opposed to the potential for harm, moving away from the computational imaginary of environments as autonomous systems. Rather, the concepts of fragility and antifragility keep the human in the loop—human dreams, human love, human allies. Antifragility imagines the possibilities for change in a positive rather than a deleterious direction. "The resilient resists shocks and stays the same; the antifragile gets better. . . . Antifragility has the singular property of allowing us to deal with the unknown, to do things without understanding them—and do them well."[90] To create an antifragile system is not simply to think of the fitness of seafloor ecosystems, but about including deliberate efforts on the part of humans to protect nonhuman environments. Like precaution, it is a way of acting in the face of uncertainty, without the hubris of assuming that uncertainty can be tamed. There is, perhaps, a place for plume research in the project of antifragility. The continued development of a plumate scopic regime in science and popular culture can play a role in connecting the biologic and geologic world, as well as human interests and more-than-human geographies.

Turbulent Transfigurations

I want to close this chapter with a suggestion that beyond regulatory assessments of resilience and risk, returning to the plume as an elemental figure for thought can help us reshape the terms by which we think about precautionary discourse. Plumes circulate in radically indeterminate ways and across multiple scales. Plume researchers talk about momentum, concentration, and entrainment, or the transport of material across an interface between two fluids on the outer edges of plume formations. These processes offer helpful analogies for thinking through the environmental impact discourse around seabed mining, as well as attending to environmental context, timing, and alignments between audiences.

Consider, for example, that the fast-paced development of the deep-sea mining industry is being driven by a momentum to invest in an era of increasingly insecure terrestrial resources. The *concentration* of this interest varies across geographies and according to political necessity. As plumes move away from their source, their turbulence diminishes. Similarly, social perception of the hazard from deep-sea mining diminishes as we move away from the communities who are most directly affected by the industry, leading to dominant views of seabed mining as a remote, "green" endeavor that will benefit the planet. All the while, there is a continuous *entrainment* and *detrainment* of scientific data into political and economic concerns—these discourses help each other along at certain intersections of funding and deep-sea research, but also diverge as they circulate in and out of social and regulatory spaces.

In addition, turning toward the ocean's unruly movements can offer new historiographic analogies that depart from tendencies to link environmental management and history to a teleology of containment. In *Shipwreck Modernity*, Steve Mentz proposes an ecologically oriented historical model, seeking to recuperate turbulence as a historical framework: "The theoretical structures I advance eschew clean transitions for messy turbulence; these historical epochs encompass a plurality that disorients, sometimes drastically."[91] Thinking about oceanic turbulence, as Mentz does, helps us to see relations of upheaval in continuous processes of sedimentation, which can be simultaneously lethal and life-giving. Plumes also point us to the uneven mixing and spreading of environmental risk into society. They symbolize the proliferation of environmental turbulences in the age of the Anthropocene and, by extension, the age of information capitalism, which drives the deep-sea resource gold rush in the first place.

To borrow a phrase from Christina Sharpe, vent-dwelling shrimp are "living in the wake," in the precarious, disturbed flows that follow from extractive capitalism and exploitation. Sharpe writes, "To be *in* the wake is to occupy and to be occupied by the continuous and changing present of slavery's as yet unresolved unfolding."[92] The comparison between the enslavement of nature and the enslavement of human beings implied in my use of Sharpe's words is not merely a metaphor or a clever turn of phrase. Environmental humanities scholars such as Kathryn Yusoff, J. T. Roane, Justine Hosbey, and Astrida Neimanis have all highlighted the ongoing ideological and material entanglements between racial justice and environmental justice issues. I particularly like the way Neimanis opts for entanglement without equivalence: "Dinoflagellates are not *like* a black body. Analogies and equivalences are their own kind of violence, perpetuating the invisibility and pervasiveness of weather as anti-blackness. We must connect these weather systems in their common alibis of force and power."[93] Neimanis insists here that while certain "weather" systems are complex and not fungible, the same forces of anti-blackness that create violence in our terrestrial world create ecological violence in our oceans. And so, we must foster wakefulness around what is to come as we preemptively mourn and memorialize those oceanic others who will be wiped out by gestures that extend property regimes to the natural world.

In this chapter I have demonstrated how plume mediations are wielded both within ocean science and without as a way of negotiating the social boundaries of environmental risk and turbulence. Outside of anthropogenic concerns, natural plumes dictate key planetary processes and create distinct zones for life in the ocean and beyond. But when it comes to extractive industry, visualizations of plumes are providing an important means of protection for the fragile, not-yet-understood eco-systems of the deep sea and water column, and are at the center of con-temporary debates around deep-sea mining and offshore drilling. This emergent plumate scopic regime has a longer history, however, within the Western world, and in particular, plume figures have a number of insti-tutional and ideological connections to environmental policy and envi-ronmentalist movements. By examining media about underwater plumes, it is clear that plumate mediations reflect both community discourses of precaution—mitigations of the unknown—and scientific attempts to address what are seen as inevitable toxic by-products of extraction through technological and biotechnological solutions.

Seafloor sediments are composed of many things: volcanic dust, sands from the coast, shells, mineral fragments, and biological debris all float downward together and combine within the water. In their circula-tions, sediment plumes are a site of crises, of futures, and of pasts. Both amniotic and abject, cleansing and dirtying, these turbulent flows are fur-ther activated in the paratextual, cultural realm, encoded with meaning as they are systematically marked and unmarked by a variety of actors and actants. Labeled, extracted, molded, and transported, sediment plumes have become humanity's latest medium of transfiguration. Toxic plumes signal livelihoods as well as potential doom. They remind us that environ-mental harm is something we experience on both a bodily and emotional level in ways that are both seen and unseen and not easily controlled.

Thinking of turbulence and sedimentation together offers a theory of change that attends to catastrophe without precluding the possibilities for settlement and deep temporalities. In the next chapter I will build on my reflections around movement tracking, scale variance, and the scal-ability of oceanographic processes by turning to the field of marine mam-mal telemetry. Like plume research, animal tagging experiments have also disrupted human scales and made ecological ones tangible. I will con-sider a different scale of marine life as well, moving from small crusta-ceans and microorganisms back to whales. These studies further prompt us to think about how global undersea information infrastructures come into conflict with local turbulences, both human and nonhuman.

4

OCEAN PACEMAKERS
Cetacean Telemetry and Ocean Health

> A whole human history, containing centuries of victory and
> defeat, expansion and retraction, trade and hardship, with all
> the meanings of alliances and recriminations, linked back to the
> natural history of cetaceans. . . . In this world whale flesh could
> become all manner of things; whale minds could speak of the
> future, and in dying, whales made men and women powerful.
>
> —Bathseba Demuth, "What Is a Whale?
> Cetacean Value at the Bering Strait, 1848–1900"

The history of human–whale relationships is a long and complex inter-
play of interspecies exploitation and empathy in which whales have played
both the victim and the collaborator. Our enduring desire to get up close
to whales, dolphins, and other cetaceans is a testament to their privileged place in the human imagination as environmentalist symbols, a
status accorded in large part because of their perceived intelligence. Historian D. Graham Burnett proposes that at the height of the 1960s and
1970s Western environmentalist movement there was "a shift to a view of
these creatures as possessed of 'intelligence'—defined loosely as cognitive and affective abilities recognizable to human beings as sufficiently like
our own (or unlike our own, but an interesting and important manner)
to disqualify them as a prey species."[1] Indeed, building off of this observation, social anthropologist Arne Kalland has gone so far as to characterize our cetacean exceptionalism as a form of totemization, wherein
attributes of multiple whale species are combined to create a mythical
portrait of a biologically, ecologically, and culturally exceptional "super-
whale."[2] Exaggerated or not, the verdict seems to be clear: humans see
whales as smart. But what are the terms by which we value and define

this intelligence? While whales appear in this chapter, my discussion of them here will not be an encounter with whale intelligence on its own terms but rather with the computational technologies that we use to remediate whales as partners in smartness.

Smartness, like sustainability, is not a neutral label, but rather one that is woven both culturally and technologically into the fabric of our society. Technology researchers like James Bridle have delved into the values that inform, for instance, the "intelligence" of artificial intelligence. "There are many different qualities which we categorize as intelligent," he contends. "They include, but are far from limited to, the capacity for logic, comprehension, self-awareness, learning, emotional understanding, creativity, reasoning, problem-solving and planning. . . . But historically, the most significant definition of intelligence is *what humans do.*"[3] Bridle mentions the existence of a varied and rich tapestry of nonhuman intelligence, but ultimately his analysis is a critique of human-defined—particularly corporate—intelligence. This corporate version of smartness pervades the world of ocean observing today, enfolding cetaceans in the process.

In chapter 2, I discussed the impact of our underwater media technologies on cetaceans and other marine organisms. In this chapter I want to go one step further to consider what it means for us to acknowledge the participation of whales in media assemblages beyond their existence as collateral damage, as co-producers of a "smart" ocean and ultimately a "smart" planet. As such, my object of analysis is not merely the whale but a technological modernity that monitors and connects us all, human and nonhuman, through a patchwork of databases and surveillance technology. Bridle's term "technology ecology" describes this doubling of the marine mammal world through data-collecting tags that allow us to see the oceans differently, to listen to whales more closely, and to differently dictate environmental protections at sea. My approach to this oceanic technology ecology is to map out some of the ways in which tag-along mediations of whales have been informed by medical, environmentalist, and corporate cultures. How did technologies like pacemakers and smartphones influence the development of whale telemetry? How might we think about the tag and its recruitment into multiple scales of global ocean observation? To answer these questions, I examine telemetry operations from the perspective of whale individuals, whale habitats (movement ecologies), and the oceans as a whole. Through interviews with marine mammal researchers as well as archival research, I trace how whale tagging has evolved from proof-of-concept devices, which are intended to bring humans closer to marine life, toward a treatment of whale bodies themselves as autonomous oceanographic infrastructures. Put simply, I explore how a desire to observe whales has led to observing the sea with whales.

I will begin with the initial motivations for the development of underwater radio telemetry and satellite telemetry under the shadow of the Cold War. This stage of cetacean tagging reflected technologically determinist desires to extrapolate existing electronic technologies to ocean inhabitants. It was not without controversy, as is evident in the ethical debates that emerged around benign observation in the early years of cetacean tag development. Next, in the wake of growing environmental concerns, I discuss the differential usage of both temporary "archival" tags and long-term "transmitting" tags for different scales of cetological research, including habitat modeling efforts. I then consider how whales are used as oceanographic sensing platforms, sometimes in conjunction with autonomous drones. This seascape cetology relies on a convergence of movement patterns between species and oceanographic processes. Finally, I consider how biotelemetry research might transform not simply our perceived relationship with whales but also trouble the anthropocentrism of ocean observation as a whole.

While much has been made of documentary images of whales, the science of whale telemetry has done much more than offer up individual animals as symbols. Like other animal telemetry networks, whale-tracking devices transmit multisensory information about both cetaceans and their habitats. Environmental media scholar Jennifer Gabrys explains that tracked animals become "biosensors" that "provide data about environmental conditions and changes," reframing whales as extensions of human technological agency.[4] Today, animal-borne sensing has allowed for the control and surveillance of remote and inaccessible environments, and data derived from marine mammal telemetry are used to assess and regulate areas of ecological importance. Beyond scientific aims, however, this transformation of cetacean bodies into signal platforms has important cultural implications for marine megafauna conservation efforts, shifting focus away from empathetic identification with whales and toward global data coverage, accessibility, and interoperability.

Importantly, an emergent relationship between underwater autonomous sensors and whales as coworkers in mobile ocean-observing points to a need to consider other implications of viewing whales in terms of their data potentialities. There remains a perception that tagged images serve primarily to produce emotionally poignant mediations of whales as precarious migrants. This precarity is extrapolated to the precarity of the oceans as a whole, and it is related to the fantasy of a pristine wilderness that can be "preserved" from human impact. But while these ideas of preservation and care are made possible through tagging, this visual and affective identification with whales both influences and stands at odds with military and commercial knowledge cultures that have also dictated the development of satellite and radio telemetry. Earlier forms of cetacean research tended to extrapolate transhumanist values of augmented capacity and ability onto marine megafauna—ideas that have been maintained throughout the history of tag development.

In the pages that follow I will make the case that monitoring the beating hearts of whales is an act that builds a sense of remote intimacy between marine animals and humanity, as well as between humans and the ocean's health as a whole, making tagged whales important cultural ambassadors and assets to an emerging global ocean monitoring system. However, taking into account the distinct medium affordances of the tag, this is not as simple as generalizing human empathy for whales to the oceans. The data intimacy that I will describe, which takes its cues from medical practice as well as from a regime of terrestrial dataveillance, works against assumptions that intimacy necessarily entails a mutually beneficial relation. As scholars such as Lisa Lowe have argued, the entanglements of intimacy and benefit cannot be decoupled from histories of exploitation and colonization.[5] Consequently, I understand intimacy as a condition of both precarity and possibility, an affective closeness that might include violence, exploitation, and voyeurism in addition to empathy and friendship. Tags are a form of extractive mediation in the sense that what is extracted from the whale are data. To the extent that intimate observation creates opportunities for exploitation, I argue that tags can also be recruited into other assemblages for mineral resource extraction, although, as I will demonstrate, this outcome is not inevitable. Marine mammal telemetry uniquely operates within multiple, intersecting spatiotemporal resolutions as well as multiple social worlds, and is positioned to mediate human–ocean intimacies more broadly. Although we have tended to project our own values onto whale bodies, telemetry also offers opportunities for distributed knowledge, care, and an orientation toward collective survival.

The Problem of Tracking Whales

Historically, the study of whale movements was a specialty found among whalers and mariners, who had a direct stake in being able to predict the movements of cetaceans. Hunters, whose primary products included sperm oil, spermaceti, whale bone, and whale oil, acquired specialized knowledge and skills about their quarry. In the mid-1800s and early 1900s, whaling boats recorded data about weather and whale sightings in logbooks, providing a rich archive of data that continues to be important today. At the University of Washington, for instance, the Old Weather citizen-science project recruits everyday users to go through the logs of historic commercial whaling ships to aid research on climate change.[6] Sightings on ships continue to be an important method of tracking whales today, and many of the techniques have changed little over time, such as the method of using multiple independent observers aboard a ship and then triangulating their notes to locate whales.

Even with such methods, whale behaviors, while partially known, have remained largely inaccessible to humans. We pay for boats, journey far and wide, just for a glimpse of a backfin, a tail flip, and, if we are lucky, a chance to see whales in action, breaching, feeding, communicating. In popular media about whales it is commonplace to see language about acquiring sneak peaks into the lives of these animals, whether the context for that mediation is images taken by scientists, or marine mammal theme parks that promise the public up-close views through aquarium windows and live cams.[7] Today, these once rare, intimate perspectives on whales are increasing alongside the deployment of new technologies such as underwater drones and telemetry tags, which provide a tag-along, in situ view of cetacean lives. In one recent example, a flurry of media attention arose around GoPro-like footage taken by cameras attached to whales with suction cups by the University of British Columbia's Marine Mammal Research Unit, which allow viewers to "take a ride" "as if on the back of an orca" (see Figure 10).[8] These images have brought whale social lives into greater focus, displaying the family bonds and behaviors within the tagged pod of orcas. In addition to suction cups, implant tags now allow scientists to monitor the habitats and biological functions of whales, bringing us even closer to the so-called secrets of cetaceans. Animal tagging began with bird banding in the late 1800s. However, research and development around wildlife radio telemetry started in earnest during the Cold War, on the heels of military electronics innovations in sonar as well as radio. Etienne Benson traces this history in his volume *Wired Wilderness*, which touches on work involving the Cedar Creek Natural History area in the 1960s. William Cochran, a member of this group, led the development of radio tags that would eventually popularize the use of biotelemetry worldwide. However, while tags for birds, foxes, rabbits, badgers, and other small terrestrial organisms

Figure 10. Researchers at the University of British Columbia's Marine Mammal Research Unit examine a suction-cup Customized Animal Tracking Solutions camera data logger during their Orca Quest 2020 expedition. The data logger includes a high-definition camera in addition to satellite telemetry, a hydrophone, a time-depth recorder, and other sensors. Photograph by Marine Mammal Research Unit, University of British Columbia.

developed quickly, tags for large mammals and marine organisms were slower to come to market. As Benson explains, the combination of a perceived "undersea lag" in conjunction with anxieties about the Soviet Union being ahead of the United States in studies of underwater life worked to motivate the U.S. Navy's eventual involvement in developing aquatic tags.[9]

The picture painted by Benson's account highlights the dubious ethics of tagging captive whales through partnerships with commercial institutions and the military. Benson refers to the development of cetacean telemetry in San Diego in the 1960s and 1970s, where commercial exhibitors like Sea World partnered with the U.S. Navy's Marine Mammal Program and other researchers to acquire and tag whales. The first successfully tracked whale, a gray whale raised in captivity by Sea World named Gigi, was famously tagged and released into the wild in 1972 after the costs of keeping her became untenable. Whale telemetry experiments were thus embedded not only in a lucrative aquatic entertainment industry but also in military programs such as the U.S. Navy's Deep Ops project, which attempted to train orcas to retrieve objects. Max Ritts and John Shiga detail this period of military cetology, calling attention to the role of whale research in contributing to the "inappropriate contiguities" between military and environmentalist cultures in the 1970s.[10] Moreover, many of these early whale tags did not bear out data that were readily usable to scientists, and tag developers often found themselves pressed to justify the scientific merit of their work.

While famed cetologists like William Evans, W. W. Sutherland, and Roger Payne contended with internal ethical debates and efforts to secure funding support for radio telemetry in the 1960s, other groups of researchers were simultaneously finding new entryways into whale telemetry in the 1970s and 1980s, in the Pacific Northwest and further south, in the Amazon River. In 1987 a Marine Mammal Commission workshop gathered to discuss satellite telemetry in particular and its potential to aid the protection of endangered large whales from offshore oil and gas development, fishing, and shipping—a conversation connected to the concerns around seismic blasting. In a 1988 annual report to Congress, cetacean tracking workshops sponsored by the Minerals Management Service and organized by the Marine Mammal Commission are discussed at length, concluding that "tracking radio-tagged whales from satellites offered the best potential for obtaining needed movement and related data" and "problems with attachment and retention of tags pose the greatest obstacles."[11] These discussions would go on to inform the design of implanted transmitting tags at Oregon State University, where researchers experimented with everything from suction cups, barbs, darts, shotguns, and more to develop workable tags for wild cetaceans.

To gain insight into the challenges of tagging whales as well as the history of cetacean telemetry research, I spoke to Daniel Palacios, the current lead researcher of the Whale Habitat, Ecology, and Telemetry (WHET) Lab at Oregon State University.[12] He explained that tagging techniques necessarily cater to the animal species in question. Seabirds, for instance, are caught in the nest and then banded. Elephants are caught in colonies and then fitted with a tag via an adhesive or glue. Fish are first caught on board a ship and then tagged, while shark tags are clipped to the fin. Cetaceans pose special challenges because of their shape and large size. Palacios describes some of the challenging logistics: "The way the whale's body is designed offers very little to no opportunity. It's

Figure 11. This fully implantable tag on a gray whale only records location. Image courtesy of Whale Habitat, Ecology, and Telemetry Laboratory at Oregon State University's Marine Mammal Institute. Photograph by Craig Hasylip, Marine Mammal Institute, Oregon State University

not like their flippers are accessible to the researcher, even momentarily, or longer term."[13] The streamlined body of a whale is not conducive to bands, collars, fur or feather attachments, or other conventional methods used on terrestrial organisms. Other techniques, such as glues that are used for pinnipeds' hair, do not work for cetaceans as their outer layer of skin cells is sloughed off on a regular basis.[14] These challenges are compounded by the fact that whales typically have to be tagged while in motion in the ocean.

Underwater archival tags, or temporary tags (typically using a suction cup or clamp method), were invented first. These largely analog devices initially recorded time and depth, and do not transmit information. Rather, they are meant to fall off the animal and be recovered, which makes them ideal for easily recaptured marine animals. Archival tags are commercially abundant today and carry a range of different sensors due to their relative ease of development, including underwater video cameras. Satellite transmitting tags for seagoing mammals were much trickier to develop, however, and require implant mechanisms for longer-term data collection (see Figures 11 and 12). Even when implanted,

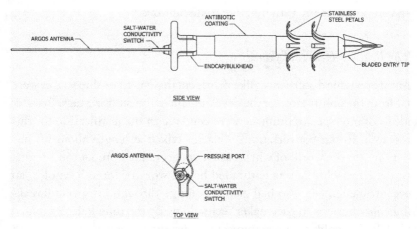

Figure 12. The implantable satellite tag design currently in use by the Oregon State University Whale Habitat, Ecology, and Telemetry Lab consists of a main body, antenna and external sensor endcap, penetrating tip, and anchoring system. Image originally published in Daniel M. Palacios et al., "A Satellite-Linked Tag for the Long-Term Monitoring of Diving Behavior in Large Whales," *Animal Biotelemetry* 10, no. 26 (2022), https://doi.org/10.1186/s40317-022-00297-9.

transmitting tags can still get lost and eventually work their way outside of a whale, much like a splinter. Other barriers to development included ethical debates both from other scientists and from the public about the extent to which tags could truly act as non-intervening forms of observation, as well as an initial lack of information sharing among researchers about how to design transmitting tags for whales.[15] Given these constraints, tagging development for whales took a relatively long time, and continues to lag behind most land animal tags. To make matters more complicated, satellite tags only transmit when above water and are limited by power and bandwidth, reducing the amount of data they can send and limiting their use to air-breathing species like whales. Palacios and his colleagues describe information sent by satellite tag as working similarly to a smoke signal, as not much information can be packaged in the transit from the ocean surface to the Argos satellites that encircle the planet. Tags are thus often pre-programmed to transmit data summaries or random samples of more detailed information.[16] The basic data are then analyzed through models which allow researchers to infer things about unobserved behaviors, eliminate errors and outliers, and smooth out any variation due to disruptions.[17]

The Beating Hearts of Giants

Anthropogenic disturbances like debris entanglement or ship strikes were far from the only reason for the development of transmitting tags. Broader ideals of interspecies intimacy were centered in the justification for this research. In our conversations, Palacios talked at length about his initiation into the world of whales as an undergraduate student in Bogotá. In the late 1980s he was enthralled by the work of Jorge Reynolds, an eccentric engineer who had gained prestige through his groundbreaking contributions to pacemaker technology. By the time Palacios joined his lab, Reynolds was economically independent and partnering with a melting pot of thinkers, including engineers, biologists, linguists, and humanists, to explore free-flowing, hypothetical questions. He began thinking about applying his knowledge of pacemakers to animals. What would it take to produce an electrocardiogram of an Amazon River dolphin? A humpback whale?

Reynolds's start in medicine dictated his journey into telemetry. Pacemakers are not unlike tags; they are portable devices that measure biological variables. The technical challenges—including attachment, miniaturization, power, and fidelity—are largely the same. Meanwhile, Reynolds was not motivated by the same set of geopolitical concerns that his U.S. colleagues might have been, but like many early radio telemetry developers, he shared in the belief that technological development should not be hindered by the requirement of a particular question or by its practical value. This was, as Palacios puts it, "science for the sake of science itself. . . . You don't always have to start a project to find a cure."[18] In interviews, Reynolds emphasized the poetic sense of kinship he felt working with both whale hearts and human hearts. Whale hearts, while obviously very large, are similar in structure to our own, and the idea of listening into the hearts of animals conveyed an ethos of care and closeness that mirrored his sentiments toward human patients. In 2015, Reynolds reflected on his career interest in the "beating hearts of giants," musing, "I think I have contributed to the whale, not as a spectacle, but as a creature which lives for the benefit of mankind."[19] To our ears there is a glaring anthropocentrism to this statement. However, the idea of "benefit" here remains open and expansive. Reynolds's judgment against media spectacle is suggestive in the way that he posits some mediations as more valuable than others. The framing of the article reinforces the idea that unlike popular visual media, tags and pacemakers allow us to understand, on an intimate level, phenomena that would otherwise be invisible.

Reynolds's work on whale hearts failed to pique the interest of the Colombian government, which, he complained, "thinks science research is not a serious endeavor," but it was backed by other scientists and by international corporations.[20] Eventually, his interest in tagging would put him into conversations with U.S. marine mammal telemetry researchers at Oregon State University, led by Bruce Mate. There, Mate was developing the first satellite telemetry tags for whales, and he would go on to successfully use this technology to track the dwindling populations of bowhead whales off the Pacific Coast.[21] Inspired by Mate's presentation, Reynolds formed a team of industrial design students, electronics engineers, and biologists, affectionately dubbing them the "Whale Heart

Satellite Tracking" (WHST) team. Although the group was formed in 1990, it took nearly thirty years for their goals to come to fruition, with the first electrocardiogram of a free-ranging humpback whale done through satellite tagging in 2019.

The Scale of a Whale

Data from tagged whales provide information to researchers at a variety of temporal and spatial scales of observation. Questions about discrete whale behaviors like feeding or mating can be resolved through retrievable archival tags like those used by the whale research and conservation group Ocean Alliance. Longer-term and more geographically expansive queries—for instance, about migrating whales—require implant tags that can stick around longer. Given the different kinds of tags available, research questions today include the development of place-based knowledge around biodiverse areas.

Early radio telemetry sought to excise much of the environment from tag data, as environmental factors contributed to signal bounce problems that impeded information about the animals. Now, the environment is very much centered in our understanding of species. The effect is that animal mobility is no longer a problem to be resolved as it once was in the era of whaling, but rather, movement is an asset to the datafication of the oceans as a whole. Movement ecology research addresses fluctuations in populations, environmentally important areas, and animal responses to environmental shifts (such as climate change).[22] Gabrys explains, "If animals are on the move, so too is ecology adopting a more mobile method in order to understand these questions."[23] The movements of populations are also part of what dictate the scale of observation, particularly if the collected data come in the form of GPS coordinates and CTD (conductivity, temperature, and depth) measurements. Often, decisions about what kind of technology to use depends on the species of whale, as different populations move through ocean space differently; some are seasonal visitors and migrants, whereas others are full-time residents.

The other piece determining the scale of telemetry data collection involves the environment itself and the temporality of processes that are specific to a given location. Elizabeth Becker, a NOAA fisheries researcher specializing in remote data and habitat modeling, explains, "You have to look at the spatiotemporal resolution that the environment is changing on."[24] Waters of the U.S. West Coast, for instance, are extremely dynamic systems that include upwelling and El Niño processes, requiring observation on a weekly, annual, and seasonal scale. The production of open-ocean currents is the result of the Coriolis effect, wherein the Earth's rotation generates a force perpendicular to the motion, thereby deflecting circulating winds toward the poles. These currents in turn shape the coastline and seafloor as well as transfer heat, nutrients, and pollutants. It is this process that creates ocean provinces—distinct habitats and environments.[25] Jessica Lehman describes the production of planetary knowledge through the "synoptic geographies" that it creates, defined as "a set of coordinated data practices" that operate inductively by "linking distant places through careful coordination to produce coherent and quantifiable understandings of the Earth as a planet."[26] Such a view contains both grounded and abstract perspectives of the planet. Tagged animals can provide that grounded view, while also acting as nodes that contribute to the expansion of a larger signal footprint made up of multiple movement ecologies, from which further queries into the data can be made.

All of this is to say that whale migrations offer researchers much more than merely access to cetacean lives. The object of our identification is no longer the technologically augmented whale, but a wired ocean. Collection of environmental data through telemetry informs habitat models, which have a range of oceanographic and industrial applications. On the one hand, these networks can bring attention to the need for protection in areas key to whale livelihoods, as well as to the tangible impacts that human activities like shipping or drilling have on specific species, bolstering cases for environmental protection under the Endangered Species Act. At the same time, habitat models have industrial applications. For instance, a company that is building an offshore wind energy system would need to site its project and conduct environmental impact assessments to mitigate its effect on wildlife. Using a habitat model, that company could make use of a cetacean forecast for operational logistics. And beyond these applications for building underwater infrastructure, habitat models work by resolving scales, connecting individual migrant data to much larger questions of physical oceanography.

Long-term telemetry projects in particular connect the local and the global through knowledge of whales' summer and winter homes.

Incorporated into large regional or global networks, marine animals are understood to be more than mere individuals swimming in a vast aquarium, but rather beings that are part of and dependent on their milieus. Gabrys discusses how mobile and ongoing data collections transform animals into sensor nodes that address unanswered ecological questions, "while also creating an expansive and even global animal-sensor network that functions as the 'pulse of the living planet.'"[27] Her invocation of a "pulse" of the living planet is language that comes directly from a project called the International Cooperation for Animal Research Using Space (ICARUS), which mobilizes the Argos satellite system for tracking animals. For ICARUS, the beating hearts of giants connect human empathy for the living to the more-than-living "heart" or health of the Earth as a whole. And while cetacean tagging most directly resembles the material forms of medical media like pacemakers or stethoscopes, the idea of ocean observation as monitoring health extends beyond telemetry. Others have also invoked similar medical analogies that connect high-tech ocean observation to environmental health. For instance, cabled ocean observatories (which I will discuss in chapter 5), composed of systems of networked platforms at the seafloor, have also been said to "take the pulse and vital signs" of the ocean.[28] And indeed, the metaphor of the ocean pacemaker continues to work at this larger scale: the value of ICARUS is in its ability to broadcast and network information across space, just as the central nervous system sends electrical signals to the heart.

It is one thing to compare marine wildlife telemetry to the knowledge produced by a pacemaker or stethoscope, and another to project a pulse onto the planet, as this moves us from the realm of epistemology to ontology. Invocations of metabolic processes and pulses to describe a panoply of volumetric media technologies, from space to the deep sea, point to an increasingly common perception of ocean scientists as ocean doctors and of the ocean itself as an ailing patient. Ironically, metabolic imaginaries are also connected to a Marxist political economy of production through the idea of "metabolic rift," which connotes the separation or alienation of humans from nature by labor.[29] Situated in this context, it is striking that the use of the pulse analogy to describe ICARUS is intended to have the opposite effect. Ocean pacemakers direct us to care about the ocean and about cetaceans in the same way that we care for human patients, connecting our beating hearts to the "beating hearts" of much larger bodies—whether cetacean or celestial. Rather than a metabolic rift, ocean networks are positioned as a form of metabolic suture.

Fundamentally, however, a projection of a doctor-patient relationship onto a wild environment is premised on the extrapolation of human maladies onto nonhuman objects. This anthropomorphism is both a strength and a weakness of the "pulse" metaphor. It raises questions like, how do we decide which rhythms constitute a pulse of the ocean? Is this pulse exclusive to marine mammals and megafauna, or does it include plants and nonliving nature? As media objects, the metaphor, the model, and the tag itself all play a consequential role in dictating which marine matters constitute this so-called oceanic pulse, and more broadly, they draw boundaries around what matters to us about the oceans. Although this undoubtedly has a material impact on conservation efforts, a new materialist might argue, by contrast, that ocean ecologies also exceed the representational containment of both metaphor and model.

Environmental uses of satellites and tag networks often have the same kinds of pretensions to empirical knowability as apparatuses like the X-ray or ultrasound. Jody Berland draws a throughline between medical imaging and the imaging of the whole Earth as a body, arguing that satellite imaging encourages viewers to see elements of the planet as a body and to identify changes in that body as a sign of the planet's health. She explains, "This rendering-into-image is closely linked to processes of medicalization performed by modern science, particularly on the woman's body, which paradoxically disappears behind its contents."[30] Berland's argument is that these comparisons to medical technology are nonneutral choices that contribute to erasures (the woman's body) as well as the designation of value (the fetus). A critical media perspective on tag networks would similarly assess the choices made around what kinds of information are included and are therefore valued by the media platform. Just as the Reynolds lab saw the whale pacemaker as making the whale's heart beat "for the benefit of mankind," whale movements now also benefit humanity's common interests, whether this entails collecting data for climate models or creating cetacean forecasts for industry.

As I will discuss next, the relationship between ocean doctor and ocean patient is further problematized by the explicit connections between ocean networks like ICARUS and computational media. Environmentalist media like animal telemetry tags are mediated by the cultural techniques of both medicine and digital surveillance technology, resulting in distinctly data-oriented notions of care and intimacy. The prioritizations of human concerns in cetacean telemetry suggests that inevitably, the "whaleness" of that data, along with the agential capacities of the whale, will not be entirely archivable.

A "Smart Ocean"

In addition to poetic invocations of hearts and pulses, justifications for networked ocean observation are often filtered and made legible to a broad audience through references to the Internet of Things.[31] Our wearables and our phones give rise to data bodies that can be monetized, inputted into databases, and monitored from afar. Our bodies, like the bodies of whales and the body of the ocean as a whole, are now subject to what Eugene Thacker calls "biological exchange," "the ability to render the biological not only as information, but as mobile, distributive, networked information."[32] This implies a rhizomatically distributed notion of agency. As such, to see a telemetry tag as operating beyond the tagged individual shifts signification from fitness as a measure of individual bodily health to population optimization as it is managed through distributed, algorithmic systems that reflect the global ambitions of distributed sensing infrastructures.[33] Consequently, while the totemization of whales helped to inspire cetacean telemetry advancements, the image of the totemized whale is ultimately replaced with a more abstractable notion of data extraction points.

Zooming out further, marine animal telemetry is just one of many components in what has been branded an Oceans 2.0 or "smart" global ocean initiative comprising a "vast, heterogeneous, and always changing network of ocean observations."[34] Such a framework sees value in comprehensive spatial and temporal coverage and thereby privileges the global in order to justify financial support for further infrastructural development. Referring to technologies such as computer-mediated phones, houses, cars, classrooms, and cities, digital theorists Orit Halpern, Robert

Mitchell, and Bernard Dionysius Geoghegan speak of "smartness" as a logic of "geographic abstraction, detachment, and exemption" that relies on the continuous production and incorporation of data into a global system. They call this the "smartness mandate."[35] Their case studies of smartness include a broad range of digital technologies, from home appliances to electrical grids, all of which are linked through a shared logic of optimizing and making resilient a system of governing populations within technological zones.

Responding to the smartness mandate, the planetary ambitions of ocean observing likewise remediate the ocean's geographies through zoning and through the modulation of time. Indeed, advocates of animal telemetry networks emphasize the potential of this technology to facilitate long-term environmental monitoring and greater coverage of unknown areas: "Monitoring marine species is valuable not only in terms of increasing the perceived value of protected and exploited resources and minimizing human impacts, but also for the data those species deliver as roving reporters about the oceans, our changing climate and by extension our terrestrial weather."[36] As "roving reporters," animals travel to regions that other technologies and species often do not, including areas beneath permanent sea ice in polar regions, or remote atolls.[37] Increasing coverage was a major point of emphasis among marine megafauna researchers at the decadal OceanObs'19 conference. For instance, Ana Sequeira, a telemetry researcher at the University of Western Australia, emphasized that megafauna are capable of "increasing spatial and temporal coverage" through the incorporation of telemetry data into operational ocean-observing systems. Mark Hindell, a seal telemetry and climate researcher at the University of Tasmania, echoed these sentiments in his own talk at OceanObs'19, in which he touted biotelemetry as a tool to "help us observe the unobservable."[38]

Reflecting on these calls for coverage, I am reminded of Alberti's velo (veil) as described by Anne Friedberg: a framed, gridlike regime of visuality that was used to map three-dimensional worlds onto two-dimensional planes.[39] As a manifestation of Alberti's velo, the grid is a distinct cultural technique of ordering and addressability that can be linked back to the history of ocean navigation. Bernhard Siegert traces a throughline between maritime coordinate systems, "the mesh formed by latitudes and longitudes," to colonial regimes like the township grid, describing gridding as a media operation geared toward representing humans and things for governance.[40] Meanwhile, Geoghegan echoes Siegert's framework of cultural techniques in his discussions of how smart technologies render environments. In particular, he posits that computational media endow unruly spaces with a sense of conformity and stability through technical addressability and global standards, thereby taking part in "a political ordering of place, users, images, instruments, and space."[41]

Drawing on these theorizations of grids, we can speak of telemetry networks as participating in a similar kind of "veiling" (albeit a more organic meshwork than a straitlaced grid) through their use of location data. At the same time, comments by marine telemetry users around the need to increase data coverage speak to the way scientists also think of natural surfaces as already "veiled," while imaging technologies (borrowing Berland's term here) "pierce" and "unveil" their topographies.[42] To position technological mediations as participating in both veiling and unveiling nature heightens the agency endowed to the observer as the landscape is transformed through observation. In other words, the blending of calls to increase coverage (or to put it in Siegert's terms, to veil for the purposes of ruling) with the idea of unveiling secrets of nature doubles the anthropocentric premise of ocean observation, linking it to histories of colonial governance and to a utilitarian environmentalism wherein humanity is given a sense of control over nature. Directional sense, wayfinding, and embodied spatial relations that surpass the binaries between "relative" and "absolute" space are increasingly replaced with this new ontological order, which gives primacy to technologically distributed sensory regimes. As computational technologies spread and are made perpetually mobile, perceptual capacities are also increasingly fixed and standardized through a form of technological assimilation.

Nevertheless, while the ambitions for tagging are vast, limitations still stand in the way of this desired technological ubiquity. Satellite tags are notoriously less effective at depth, extending our existing cultural associations between depth, darkness, and the absence of knowledge. Members of the WHET lab recently discovered that dramatic changes in temperatures at depth increases errors in location accuracy, and tags have yet to evolve to compensate for this effect.[43] The paucity of telemetry data about the deep ocean and the relative scarcity of benthic infrastructures means that habitat modeling is also less reliable at depth. "Our models for the deep divers have performed the worst," explains Becker.[44] There are several possible explanations for this. As the amount of data decreases, more of it must be bundled together to be fed into models, leading to more assumptions about that data. For instance, sperm whales, despite traveling in different groups with different behaviors, are treated as one species. Oceanographic conditions factor in as well: "We don't have a good variable to capture what is important to these animals. We thought it might be the oxygen minimum layer, because that's what the squid are affected by." Becker's research has yet to identify an oceanographic variable that correlates closely enough to deep-diving whales and their prey for effective modeling. This means that the fields of marine biology and physical oceanography increasingly overlap in their research questions, as models derived from telemetry depend on a holistic understanding of ocean ecosystems and oceanographic processes.

An outcome of all of this is that our understanding of whale movements relies on probabilities, correlative models, and locative data rather than discrete sensory knowledge, signaling the most likely trajectories of each species.[45] This alters how we define intimacy; it is no longer a matter of physical or emotional proximity, but of data quantity, coverage, and precision—the asymptotic approach of speculative ocean mediations to an "absolute" nature. This shift toward probabilistic thinking enables larger-scale questioning, but it also exacerbates asymmetrical relationships between researchers and tag carriers. In the next section I will elaborate on how tagged whales and autonomous sensing platforms are recruited into a geontopolitical regime of power.

The Whale and the Drone

On the subject of radio-tagging migrating birds, Cochran once reflected, "You see yourself interfering with his life. Then he's gone. You don't think of him as a living being. He takes on a different place in your mind. He becomes a data machine."[46] Tracking remediates its animal hosts by "rendering their bodies into signals—intelligible, decipherable, traceable—while also compressing the immense distances they cross into the two-dimensional coordinates of a screen or a digital map."[47] As emotional intimacy is replaced by ecological addressability and dematerialized data access, the gaze of the researcher is disembodied as well. What Cochran intuited about this dynamic is that a cartographic view often excises the question of empathy between multiple individual whales as well as the feelings of interspecies kinship that initially drove people like Reynolds to record the beating hearts of whales. This decoupling of intimacy from kinship suggests the need to question assumptions about intimacy that cast it as a monolithic category of experience that implies care. What we see in biotelemetry developments is rather a multiplicity of intimate relations, with differing attitudes, affects, and attachments.

Tendencies to abstract or separate networked ocean observation from the realm of social and cultural experience are not only an outcome of technological limits, the filtering of data collection to a few specific variables, but also reflect a longer history of utilitarian thinking about nature in Western environmental research. One could argue that the objectification of whale bodies as sensing machines starts during World War II, with U.S. naval explorations of echolocation and undersea sound. During this time, naval researchers took note of the unique sonar-based communication of whales, and used those studies to construct mechanized systems of submarine communication. As Ritts and Shiga explain, "Listening to echolocation marked an evolution in the acoustemology of naval cetacean research. It elevated the significance of cetacean sound production and enriched the analogy between cetacean bodies and submarines."[48] This was accompanied by reconfigurations of animal behavior through the lens of logistics: target acquisition, optimization, and budgeting were all concepts used to describe whale behavior, thus transferring military knowledge cultures into the realm of cetology and marine biology. Meanwhile, autonomous underwater vehicles, developed in the 1980s, were often made to resemble torpedoes (a functional design choice that helped attract military funding), and torpedoes in turn were modeled after the torsos of dolphins.[49]

While they seldom receive the credit today, studies of fish and shrimp sounds were also crucial to the understanding and militarization of underwater sound. Marine biologist Marie Poland Fish's groundbreaking bioacoustics work on Atlantic fish species was, for instance, funded by the U.S. Navy and used to help train listeners during the Cold War. More recently, the sounds made by snapping shrimp are being reclassified as "natural sonar" and are among several biologically produced sounds that are being vetted by the Defense Advanced Research Projects Agency for an underwater submarine detection system.[50] Previously, I referred to this as a system of ichthyoveillance, noting how in official media communications about the Persistent Aquatic Living Sensors program (PALS), animal subjectivity is stripped away for discrete functionality, reconfiguring fishes as networks for signal detection.[51]

Ritts and Shiga posit that such ideas reflected the rise of bionics, or the "scientific fusion of biology and engineering to optimize or enhance bodies with sensory prostheses."[52] Indeed, it is tempting to think of tagged whales as bionic or cyborg specimens enhanced through prosthesis—especially when it comes to early experiments such as the Reynolds whale pacemaker. However, my conversations with contemporary biotelemetry users suggest that the stakes are larger than the bionic whale; scientists are excited about the enhancement of a much bigger sensory infrastructure with the data collected by whales. This tracks with the distinction between bionics and cybernetics. If we can speak of a "posthumanism" with the cybernetic era, this raises the question of whether telemetry also participates in a "postnatural" age in which all animals are fundamentally transformed by cybernetic structures.[53]

Both military technology and cetacean telemetry are driven by desires to expand human knowledge and control into inaccessible areas, and this is what ultimately links whales to drones. Consider this justification for animal-borne sensors (ABS) by its users:

> ABS are mobile, autonomous platforms that are relatively inexpensive to deploy (compared to gliders or AUVs), provide important insights into US coastal and EEZ areas, and are particularly useful in the open oceans that are difficult and expensive to monitor. . . . Animals are adept at finding areas of particular interest to oceanographers, including surface and sub-surface fronts, eddies, and confluences that aggregate prey.[54]

Here, marine megafauna are explicitly compared to drones (gliders and AUVs) and are framed in terms of their utility to researchers. Meanwhile, justifications for the development of unmanned drones are often very similar to those made about biotelemetry. Take, for instance, the language used by NOAA about surface drones: "Unmanned gliders sample the ocean in places where it impractical to send people, and at a fraction of the cost. . . . It is these characteristics along with advancements in sensor technologies that make gliders increasingly important as tools for collecting ocean data."[55] This passage echoes the language used for animal-borne sensors, invoking both the capacity to monitor remote areas as well as a desire to keep costs down.

This rhetoric of increasing ocean coverage with tagged animals and autonomous gliders essentializes ocean wilderness by positing the technologization of nature as the solution to planetary health. But adopting this perspective, whether for the good of whales or the good of humanity, ignores the premise of control embedded within this technologization. As telemetry observers increasingly act as stalkers occupying a version of the disembodied, God's-eye view, the objects of their gaze—migrant animals—become biopolitical subjects that must be controlled through their visibility, constituting a panoptic power relation.

Michel Foucault defines panopticism as power that immobilizes its objects, but is itself made mobile through lightweight, invisible, and efficient design, not unlike attempts to make tags smaller, more durable, and more geographically comprehensive.[56] Other scholars such as Hugo Reinert have described the "asymmetrical intimacy" of telemetry as a *"constitutive withdrawal:* not an abandonment but rather the opposite, a powerful and highly productive investment which, through its paradoxical absent-presence, offered a solution (of sorts) to the double bind," engagement through absence.[57] As Reinert notes, the development of telemetry has progressively expanded the capabilities of observation technology, including real-time monitoring and increased battery efficiency, while minimizing their material footprints through miniaturization.[58] The Oceans 2.0 vision therefore posits that the mandate to colonize all the world's oceans with computational, networked technology will obviate the need for more invasive forms of environmental mediation. In doing so, ocean-observation networks maintain a certain fantasy of untouched nature within their postnatural transformation. Despite a long history of Oceans 2.0 infrastructure physically altering marine landscapes and lives, tags participate in a return to antiquated preservationist ideologies and imaginaries of a "pristine" wilderness that is separate from man.

The march toward efficiency and miniaturization in telemetry is in large part influenced by human-oriented technologies like the cell phone, which provide the template for mobile surveillance and a digitally mediated control society.[59] But while marine biotelemetry tags resemble the "liquid" power that defines terrestrial societies of control, there is an important caveat: What happens when a control society has to account for Earth systems? This is the question Lehman poses, and her answer is that in order to account for the interweaving of living and nonliving systems, power must also be conceived as operating beyond the confines of life. The type of data and research that relies on whale telemetry tends to collapse biologic and oceanographic processes, framing whales in terms of physical variables like temperature, current patterns, and oxygen levels and vice versa. As such, it is not sufficient to simply think of the Foucauldian framework for "making live" or "letting die." Lehman proposes that Elizabeth Povinelli's geontopolitics is the better option for understanding the power dynamics of ocean observation, because it allows for a focus on potentiality and power: "Potentiality can no longer be defined, even ambivalently, as the provenance of life, but rather as something that characterizes nonlife entities such as Earth systems."[60] Indeed, tagging equates its living hosts to nonliving observation platforms, privileging the potentialities latent in whale movements over the value of cetacean lives.

Boundaries between life and nonlife are not easily reinstated, despite tendencies to narratively reproduce this boundary as endemic to the work of marine science, natural history, and resource extraction (see chapter 1). Life in the tag assemblage ceases to be a binary concept that is tied to biological death, but rather enters into new realms of signification with the added connotations of signal transmission. Whale life becomes entangled into the "life" of the signal, and those two heartbeats do not always overlap. As Alexandra Palmer explains, several configurations of life and death are possible between tags and animals. At times, tags may stop transmitting even while still attached to the whale, resulting in signal death that "can render animals' burden purposeless" and stand in for animal death.[61] Drone researcher Adam Fish explains: "Like the species they observe, these drones often die a death of destruction and disrepair."[62] Meanwhile, tagged whales may themselves die while their tags remain functional. Just as whale lives exceed the functions of the telemetry assemblage, transmitting tags also exceed the labor they are designed for. If we are to see the tagged whale as an autonomous sensing platform, then this whale-as-drone must itself be understood as a subject with a lifespan that is more than biological. Drones mirror tagged whales in more ways than one. For example, there has been growing interest in the development of marine drone fleets, thanks to companies such as Saildrone which are able to produce wind- and solar-powered gliders at low cost. Echoing desires to increase the geographical coverage of ocean observation, the company's slogan is "Any Sensor. Anytime. Anywhere." I spoke to an engineer at Saildrone, Dillon McEwan, about the perceived benefits of such fleets. McEwan explains that in addition to being able to measure variables like CO_2, sound, or eDNA, drone fleets can detect and follow tagged animals, and are used for operations that last two to three months. Like transmitting tags, however, these drones are tied to the surface through their power sources as well as by satellite transmission. Saildrones are also AI enabled, which means they are capable of machine learning analysis on images of animals. Having drones and gliders trail tagged animals is therefore useful in that they can pick up additional information about the environment and the animal population.

Underwater drones that can follow and potentially film whale pods expand existing relationships between drone photography and environmental imaging. Aerial drones have already been lauded for their abilities to provide rare sightings of elusive cetaceans such as beaked whales, the deepest diving mammals in the world.[63] The prospect of in situ drone footage could only extend the existing appetite for a glimpse into the "hidden lives" of these animals. Already, Saildrones used by marine mammal researchers at NOAA to study the diet and health of fur seals have been lauded for providing a seal's perspective. In one of the team's dispatches from the field, ecologist Carey Kuhn writes about the combination of tags and Saildrones: "These instruments record high-definition video, which will allow us to take a trip to sea from the point of view of a fur seal and record each time they feed. . . . By linking the video data with the Saildrones echosounder data we can determine if feeding rates vary relative to the amount of fish that's available."[64] There are other drivers for glider technology as well. Beyond following marine animals, Saildrones have applications in industry, such as conducting bathymetry surveys for wind turbines, mapping emerging channels and shorelines as arctic ice melts, and conducting offshore energy research. The same excitement over this ability to remotely observe the most intimate

Figure 13. Saildrone exhibited at OceanObs'19. Photograph by author.

workings of whale bodies and previously unseen behaviors apparent in the WHST team is reflected in recent innovations around autonomous vehicles and tags. In 2022 a group of cetacean researchers with the Ocean Alliance succeeded in using drones to remotely attach four suction cup tags to a blue whale (see Figure 14). In a blog post announcing the success, researcher Iain Kerr writes, "Think about that, whale surfaces, drone launches, flies over the whale, drops the tag and flies back to the boat . . . in less than two minutes!" Kerr goes on to describe the method as a "potentially revolutionary research tool," reflecting an enthusiasm over the ability to film and collect data about whale feeding behaviors without the stressors of pole attachment methods.[65] The team's original purpose revolves around the SnotBot, a drone designed to follow and collect whale blow exhalations and analyze the blow for DNA, hormones, and other biological information.

While the Saildrone and the SnotBot may be some of the first drones to follow and collect data about marine animals, they will not be the last. At MIT's Computer Science and Artificial Intelligence Lab as well as the WHOI Autonomous Robotics and Perception Lab, researchers are working on building underwater drones that will one day eliminate the need for tags altogether. This includes charismatic bots like the Robo-starfish,

Figure 14. Ocean Alliance was one of the first groups to successfully attach a Customized Animal Tracking Solutions tag on a whale remotely using a drone, as pictured here on its 2020 SnotBot expedition in Baja, Mexico. Photograph by Ocean Alliance.

a silicon creature designed to passively interact with the fluid environment and stealthily measure data in the deep sea without disturbing marine life.[66] Like the Saildrone, many of these animal-tracking drones use artificial intelligence and would be able to gather information and identify targets on the fly using real-time vision-based methods of tracking. The military or carceral applications of intelligent, animal-seeking autonomous machines hardly needs to be articulated. One ocean engineer I spoke to talked about how he was already feeling uneasy about the potential social outcomes of his research and the fact that it could be weaponized in the human world. Thus, the excitement over remote intimacy is shadowed by another, sinister application—the power to not just record but also to end a pulse.

As I reflected on the techno-optimism of wildlife telemetry and autonomous drone development in our oceans, I began pondering the stakes of what many marine mammal researchers see as the ultimate goal of biotelemetry: to create a global, one-stop shop for ocean data—an aquatic Amazon, so to speak, wherein a large range of information about marine megafauna and their habitats is freely accessible and available to researchers in standardized, integrated formats. Researchers like Sequeira and Mike Fedack have called for much larger sample sizes of tagged animals, along with globally integrated databases of marine megafauna observations.[67] While there is a democratic, caring premise to this notion of ocean pacemakers, the digital fetishism behind drone fleets and real-time monitoring of wildlife often simplifies or ignores the material lives and deaths of these technological assemblages, and limits the options for alternative modes of conservation and care. There are also other problems to consider. The aquatic Amazon idea ignores instances in which sensor data are considered an issue of national security or military interest, leading to limitations in researcher access to data. These technologies could also worsen problems with poaching and illegal hunting, making whales easier to find.

Fish reiterates these critiques, specifically questioning the notion that drones could be the solution to all manner of environmental issues, including extinction, poaching, and environmental justice.[68] He argues that this fosters a dependency between conservation and remote sensing, creating a "trap with humans and their technologies of governance, policing, surveillance, management, and protection."[69] Networked ocean observation has become just one instance of a widespread fantasy of what Gabrys describes as a programmable earth, wherein the ecosphere is imagined to follow and respond to the technosphere. As numerous environmental humanities scholars have pointed out, however, this relationality is diametrically opposed to the material realities of how digital technologies are produced and used, including their reliance on natural resources and their contribution to waste in the ocean. After all, tags are still material objects that interface with the environment, and can add to the ocean's junkspace in addition to leaking toxic chemicals from their batteries.[70] This particular problem is being acknowledged to an extent, however; future innovations in tagging will likely seek out biocompatible materials as well as renewable sources of energy for transmitters.[71]

Although there are biases and limitations to the use of medical metaphors in oceanography, we can also find purchase in their critical affordances. The UN refers to the oceans as the lungs of the world because they produce most of the oxygen on the Earth via seaweed and other aquatic plants.[72] Perhaps, then, the deployment of animal telemetry networks discussed here can be likened to intubation, the insertion of medical tubes to help living bodies breathe. While the analogy may be imperfect, it acknowledges that these distributed data feeds signal a codependency between the ocean and human technologies. This implies a major shift in how we organize and interpret knowledge about the ocean, how we imagine ourselves in relation to deep-sea environments, and how we hope to avoid extinction. Today, what we are dealing with is not just a sick ocean or a new frontier to be conquered, but the prospect of a permanently intubated patient. Given that marine animals are part of this medical apparatus, they become essential workers as well, as exploitable as any Amazon employee. Ultimately, one cannot ignore the artifacts of colonialism and militarism within these conservation technologies.

The Mentorship of Whales

For much of this chapter I have outlined ways in which we transform our perceptions of whales through tagging mechanisms. But equally important is how these technologies transform and decenter human observers. For all their limitations, tags have changed our understanding of the ecological relationships between species and their environments, as well as troubled our assumptions of what animal intelligence and socialization looks like.

Recently, whale researcher Briana Abrahms partnered with Palacios and the WHET Lab to track blue whale migrations in relation to phytoplankton blooms. This study was intended to test whether or not blue whales "surf the green wave," timing their migrations with resource waves in order to maximize their gain. Synthesizing ten years of satellite tracking data alongside simultaneous oceanographic variables in the North Pacific, the authors found evidence that long-distance migrations of whales are informed by resource tracking. This discovery simultaneously transformed our understanding of whale memory: "We find that blue whales surf climatological resource waves, using memory to track shifting hotspots of predictable and high-quality resources. Long-term memory has been shown to be a strong driver of migration patterns across taxa and in some cases a stronger driver than proximate cues."[73] Studies like this upend existing human understandings of memory, movement, and sustenance. They also generate new questions for human-environmental relation: What if we were to taking surfing as a conceptual basis for our relationship to environmental resources, rather than extraction? What lessons might the long-term memory of a whale have for researchers working to model and forecast environmental conditions through technological means? These insights work against dominant tendencies to fixate on human exceptionalism, biological limitation, and technological augmentation—ideas that lead to notions of whales as human-like and humans as intrepid underwater pioneers. By contrast, studies of whale memory and whale perception dismantle hierarchies of knowledge and sense, verging toward novel understandings of distributed sense, intelligence, and agency.[74]

Perspectives on whale networks also tend to ignore the fact that the existence of whales as networked beings is not limited to their technological enhancement. What does a distributed network look like with empathy added back in? All one has to do to answer this question is look to the whales themselves. Whales are highly social organisms, possessing global communication networks that are interwoven with their capacities for empathy. Neuroscientist and cetacean expert Lori Marino posits that the large paralimbic systems of cetacean brains are likely used for complex emotional processing and social cohesion on a level that exceeds human experience: "It suggests there may be something about their sense of self that is not only an individual, but depends very much on their social group." This distributed subjectivity, in turn, may account for behaviors such as mass strandings, which are not observed in other terrestrial herd animals. "Dolphins and whales will stand by each other. They won't leave when there's a chance to escape."[75] Reducing ideas of networked subjectivity to digital technology therefore both ignores the existence of biological networks and erases their historical role in the development of human communication networks like sonar and radio.

Whale subjectivities, which fuse empathy and intelligence, both contribute to and challenge the anthropocentrism that lies at the core of the Western world's idolization of whales. But I do not see these relationships between humans and whales as reducible to critiques that skewer the totemization of whales as a money-making scheme among Western conservationists. Marine intimacies can also expand human attentions and sensations. I think of texts like Alexis Pauline Gumbs's *Undrowned*, which derives inspiration from the lives of whales, dolphins, and porpoises in order to think through Black feminist subjectivity. Gumbs connects with marine mammals through poetic practices of metaphorization, allegory, and even breathwork:

> My hope, my grand poetic intervention here is to move from identification, also known as that process through which we say what is what, like which dolphin is that over there and what are its properties, to *identification*, that process through which we expand our empathy and the boundaries of who we are become more fluid, because we *identify with* the experience of someone different, maybe someone of a whole different so-called species.[76]

Human identification with whales has been undeniably effective in garnering support for conservation efforts. As Gumbs contends, this relationship of identification can be used not only to change the circumstances in which whales live but also to change humanity's perspective of itself. By submitting to the mentorship of whales, we might question, for instance, our tendencies to separate information cultures from empathy and kinship, as well as deconstruct those ever-looming binaries of the digital/biological, self/other, individual/community, nature/culture. As is plainly evident here, information can be intimate, and intimacy itself can be a site in which empire reproduces itself.

In the next chapter I will take up where tags leave off, with seafloor observatories, which have been designed to provide long-term mediations of the deep sea. Efforts around the world to build platforms at depth connected by fiber-optic cables extend smart technology to a space that has long been perceived as a gap in digitally networked observation. While these observatories serve both scientific and industrial goals, they also provide a unique venue for the blossoming of new relationships between humans and aquatic species. Like the tags that first gave us the ability to take a ride on the backs of whales, these observatories are visually compelling venues for interspecies encounter, creating unexpected intimacies that often exceed the profilmic space of in situ sensors and cameras.

5

DEEPWATER FEEDS
Futures of Seabed Media Infrastructure

> I share with the mountaineers a certain desire (or mania) to
> know high or deep places. Once asked about what I would see
> besides mud at the bottom of the sea, I replied, "What does
> one see on a mountain top except snow." Joking aside, one
> must actually go to see or send a recording system such as a
> camera or television pickup. And when we do send cameras
> down we do find many other things than mud.
>
> —Harold Edgerton, "The Trench of Puerto Rico,"
> March 15, 1960

Imagine the world twenty years into the future. Humanity, unable to
ignore the environmental calamities created by carbon emissions and
terrestrial extractions of fossil fuels, has fully turned its attentions to
developing the ocean bottom. Four thousand meters down, a landscape
once vast and largely impenetrable save for the occasional passing of an
AUV, is now dotted with turbines, mining vessels, aquaculture facilities,
and sensing platforms. Like rats in a cityscape, fish and crabs crawl over
these structures and duck beneath their dark corners, dodging lights,
nets, broken bits of metal, and other pieces of organic and inorganic
detritus that periodically kick up and circulate around the twilight zone.
A maintenance worker, encased in a round submersible, journeys to a
nearby undersea electrical outlet embedded on top of a fiber-optic cable,
plugging in a device that will measure and monitor the surrounding
environment for changes. Afterward, somewhere in a landing station
thousands of meters above, a telecommunications expert receives a message from the crew, signaling that the mission is complete. Meanwhile,
a lone bamboo coral waves its filaments in the current, the sole surviving
member of what was once a thriving deep-sea filter feeder community.

This dystopian picture of a deep-sea future may seem science fictional, but its foundations are already being laid. In fact, the speculative portrait that I painted above is not wholly mine, but borrowed. When I attended my first oceanographic conference, in 2019, I put on a VR headset created by the U.S. Department of Energy and took a virtual tour of a futuristic underwater landscape closely resembling what I just described, minus the presence of ocean wildlife like corals, fish, or crustaceans. We have already modeled an undersea future for humanity, diagrammed it, and even simulated it. Now, that fantasy of an aquatic future is actively being made into a reality.

A central aspect of humanity's conquest of the deep has been to expand media infrastructure into deep-ocean environments. In the 1960s, the terms by which we imagined controlling the ocean depths were set in part by media technologies like the television, which were at the time fast becoming a staple in the suburban American living room. Senator Claiborne Pell, writing in 1967, mused: "A television system has just entered the market which is effective down to 2000 feet below the surface of the sea; greater depth effectiveness is bound to evolve shortly."[1] Pell's aspirations of deep-sea TV may not have happened the way he envisioned, but the core of these sentiments persisted with the advent of in situ video, telepresence, and other underwater media technologies. In fact, the 1960s would prove to be a pivotal decade for ocean exploration, as it was during this period that the field of "ocean engineering" emerged in response to "the increasing intensity of ocean utilization, coupled with a corresponding increase in the application of advanced technology."[2] This integrated discipline around ocean operations led to groundbreaking underwater submersibles like the *Alvin* and *Trieste*, which would inspire oceanographers to consider the potential for manned seabed stations.

In 1964, WHOI oceanographer Paul Fye urged his peers to start building a bottom station where men would be protected from high pressures, with "an elevator to take them back up, and vehicles to explore around their new home." He set the deadline for the station at 1973, estimating the cost at twenty-five million dollars per station.[3] Fye's invocation of the deep seabed as a "new home" for man echoes decades of pseudoscientific speculation around both past and future aquatic humanisms—a hubris that, as I will demonstrate, continues to animate deep-sea media infrastructural developments today. In Fye's time there was a bias in oceanography toward embodied exploration, mirroring public fanfare about astronauts and space exploration. Oceanography was described as a "science for the whole man," and oceanographers were selected with a particular normative idea of fitness in mind—usually white and male—someone with a strong stomach and a brave face.[4] With the turn of the millennium, however, ocean engineering changed to reflect the computational era, which largely entailed removing the man from the station (although perhaps not from the surface vessel). Once again, the latest media technology set the boundaries of this fantasy. In 2004, oceanographer Alan Chave mused, "It would be a lot easier to explore the deep ocean, if we only had some electrical outlets and phone jacks on the seafloor."[5] Beyond a mere focus on technical and design feasibilities, the cabled seafloor observatories of today articulate existing terrestrial ideas of network connectivity, mapping the sociotechnological potentialities of the internet onto an oceanic future.

In order to understand the stakes of building out this seafloor infrastructure, I frame my analysis in this chapter in response to the following questions: In a global environment now defined by planetary monitoring technologies, what kind of speculation informs the recruitment of the deep sea into this data infrastructure? And following this, how might a cabled or connected deep sea reimagine human–ocean futurities? What new couplings between humanity and ocean wildlife emerge out of the production of underwater data firehoses? To research this issue, I interviewed oceanographers and engineers, examined documents related to the construction of seafloor information and communication technology (ICT), and drew from social media accounts that report on maintenance and labor efforts related to cabled observatories. One item of particular interest emerged during this research: biofouling, a term widely used in the marine industries to refer to the accumulation of marine colonies on human technologies, including organisms such as mussels, seaweeds, barnacles, bacteria, and protozoa.

In documents about seafloor observatory development, the notion of extending the internet underwater naturalizes ideas about human aquatic evolution in a way that privileges technological modernity and frames it as a universalizing force. However, looking at specific mediations of networked seafloor observation also disrupts this mythology. Terrestrial ideologies reinstate the boundaries between humans and aquatic life, while the observatories themselves create unexpected interspecies intimacies and encounters that escape the epistemic expectations of technologically mediated environmental observation and discovery. Appropriating Mary Louise Pratt's use of the concept to describe imperial frontiers, Helen Wilson helpfully develops the "contact zone" as a means of describing spaces of mutual yet unequal multispecies encounter where "wildlife photographers, producers, technical crew, surfers, different communities, and researchers grapple with water, wind, rocks, and ice, plankton, plants, and all other manner of life across uneven and always shifting relations of power and knowledge-practices."[6] I find great utility in Wilson's conceptualization of the contact zone as a particular way for the media scholar to orient away from the interface between viewer and screen, and toward the ocean itself as a site of media and cultural production.

The deepwater contact zone includes both acts of mediation and physical maintenance activities that contend with biofouling, such as the cleansing of interfaces, sensors, and other ocean media. Through examining the construction of networked seafloor observatories, I ultimately locate a tension between ideas of biofouling (unwanted pollutions between species) and bioenrichment (a mutually beneficial human-nonhuman arrangement). Situated in the uneven social and political landscapes of the Anthropocene, discourses of biofouling and bioenrichment come to a head, creating a speculative terrain for aquatic humanity that both depends on and exceeds the presence of underwater cybernetic infrastructure.

While the benthic ocean's digital infrastructure could include a plethora of underwater platforms and technologies, my analysis in the following pages will be limited to seafloor observatories that contain a specific set of characteristics as delimited by observatory engineers. Technologies referred to by experts as "cabled seafloor observatories" typically share three essential elements: power and bandwidth through optical fiber, plug-and-play capability for instruments, and regular service via human-occupied vehicles (HOVs) or remotely operated vehicles (ROVs).[7] Importantly, I have chosen not to limit my analysis to a particular intended purpose for seafloor observatories, focusing instead on technological design and capability (the same technology can be tailored to climate sensing, seismic detection, oil monitoring, or military surveillance).

I begin with a brief history of cabled ocean observatories, delving into their role in addressing the "missing deep." Next, through an examination of several pioneering observatories, I take a closer look at how individual networks have engaged with imaginaries of an aquatic human futurity. For the purposes of this chapter, I will focus on three kinds of media about cabled observatories and their relationship to human–ocean imaginaries: in situ images that are captured by observatory cameras and sensors; pamphlets and proposals that emphasize the idea of an aquatic internet; and cruises deployed to maintain, repair, and educate the public about cabled observatories.

My aim is to consider how cabled observatories, construed as near-unconditional amphibious technologies, interface with the agencies of the ocean itself—an agency that, as ocean scholar Elizabeth DeLoughrey maintains, is coming into focus in the Anthropocene.[8] As the effects of anthropogenic climate change have begun to imperil the seas, computation and automation have risen alongside resilience discourse as hegemonic solutions to dealing with an increasingly complex world under existential threat. For me, infrastructural contestations offer a valuable lens through which we can start to understand this *Homo aquaticus* 2.0 not merely as a fantasy of a highly evolved underwater society but as an inevitable outcome of cross-species contamination and exchange— of biofouling and bioenrichment—negotiated at the underwater media interface.

The Missing Deep

In the beginning of this book I presented the deep sea as a culturally, technologically, and environmentally distinct space from which to understand the question of mediation. The technologies that I subsequently charted, from seismology to modeling experiments, have helped to illuminate portions of the benthos for humanity's benefit, altering our understanding of deep spaces and our capacities to exploit its resources. Diverging from Marshall McLuhan's notion of media as "extensions of man," media scholar Jennifer Gabrys points out that sensor networks function to create novel "techno-geographies."[9] Yet today, many of these technologies remain limited in the quantity of data they are capable of collecting at the seafloor. Even as the planet is increasingly crisscrossed with high-speed systems of communication technology, the deep ocean has long stood as an obstacle toward the building of digitized, globe-spanning networks of data and information.

The Earth's most robust systems of networked ocean observation currently lie at or near the surface of the water. The most widely used global observing networks rely on surface technologies such as satellite data, wind- and solar-powered drifters, buoys, gliders, and perhaps most famously, Argo floats, an international system of battery-powered, autonomous floats that collect temperature and salinity profiles from the upper two thousand meters of the global ocean.[10] Meanwhile, the ability to pursue time-series research questions in the deep sea has been limited by the constraints of batteries and survey ship availability, the dominant mode through which oceanographic research is conducted. Deep-sea data have been conspicuously absent from much of the existing digital portals for ocean research, reflecting a lack of volumetric data and a need for long-term, standardized measurements of the deep. There are, for instance, hardly any on-site, time-series observations of hydrothermal vents, frustrating our ability to understand how they grow, evolve, and are affected by turbulent events. This is what led Scripps Institution of Oceanography researcher Karen Stocks to argue that "the deep community should leverage existing cyberinfrastructure when feasible, and develop a coordinated communication effort to address the missing deep."[11] Through the repurposing, design, and installation of fiber-optic cables as a means to power ocean sensors, scientists are at last beginning to reap the benefits of sustained deep-sea observation, which has included a more bioenriched perspective on benthic ecologies at sites like hydrothermal vents.

There are several active fronts in the development of networked systems of deep-ocean observation, and this includes both physical infrastructures and frameworks for knowledge. Responding to a call for more benthic data, international initiatives like the Deep Ocean Observing Strategy are working to connect existing systems and user communities and to redefine essential ocean variables for the deep sea, including measures of temperature, salinity, sea level, carbon, oxygen, and other climate-related data. Meanwhile, the building and maintenance of regional observatory networks is intended to "contribute observations to the global system via cabled observatory and moored measurement systems."[12] The eventual target for most of these systems is the global-scale deployment of deep-ocean platforms and sensors, beginning in areas of international interest such as the Clarion-Clipperton Zone.[13]

As scientists increase data coverage of the oceans through ocean observation networks, benthic spaces are what network theorists Alexander Galloway and Eugene Thacker might call an "exploit," a hole in existing technology through which potential change can be projected: "Informatic spaces do not bow to political pressure or influence, as social spaces do. But information spaces do have bugs and holes, a by-product of high levels of technical complexity, which make them vulnerable to penetration and change."[14] Indeed, it is this condition of dual vulnerability and potentiality that has driven international investments in deep-sea technology.[15] A digital exploit is defined in large part by its call for security, and echoing this, scientists conducting deep-ocean research are working to expand vertical observation capabilities for the dual purposes of securing the deep sea's ecosystems from threats and opening up potentialities for practical applications. As the "missing deep" is filled up with technology, new extractive relationships are formed between humans and ocean space, and between local and global ocean observers. Tracking these shifts can play an important role in understanding how the deep ocean will be used to secure our futures.

Notable Developments in Cabled Ocean Observation

Like many of the activities that take place in the deep, science and science fiction have often coincided in discussions of seafloor observation, incorporating fanciful ideas about humanity and futuristic technology. Historically speaking, however, seafloor observation developed in a piecemeal way—a direct result of locally specific scientific aims and the availability of existing technical infrastructure. I start with these technical details because they allow us to pinpoint the narrative omissions and assumptions that comprise public-facing media and marketing materials around cabled seafloor observation.

While proposals for long-term ocean observatories have existed for over forty years, the funding and implementation of these designs among scientific communities gained momentum only at the turn of this century. The Japan Meteorological Agency was one of the first innovators of this technology; it produced a cabled seafloor observing system as early as 1978 for the purposes of monitoring seismic activity.[16] This system was composed of metal wires for signal transmission. Then in 1997, the Hawai'i Undersea Geo-Observatory (HUGO), funded by the National Science Foundation, acquired a fiber-optic cable from AT&T and installed it at a depth of one thousand meters between Hawai'i and the Loihi volcano in order to study the volcano's behavior in real time.[17] In doing so, HUGO became the first observatory to use an electro-optical telecommunications cable (replacing metal wires), providing proof of concept for the current generation of cabled observatories.

From there, things moved quickly. Following HUGO was the deployment of the Hawaii-2 Observatory in 1998, which focuses on recording seismic activity. The first International Conference on Ocean Observing Systems occurred in 1999, and in 2000 the NSF approved the Ocean Observatories Initiative (OOI), an American national project to construct ocean observatories in the coastal and global ocean (see Figure 15). Commissioned in 2016, OOI's original seven arrays[18] were envisioned to connect with a transnational regional observatory network in the Pacific that included Ocean Networks Canada (ONC), which now runs two networks: the Victoria Experimental Network under the Sea (VENUS) and the North-East Pacific Time-Series Undersea Networked Experiments (NEPTUNE).[19] Each of these scientific cabled observatories powers multiple kinds of instruments, including hydrophones, 3D cameras, pressure sensors, temperature loggers, seismometers, CTD (conductivity, temperature, and depth) instruments, flow meters, fluorometers, magnetometers, oxygen sensors, turbidity meters, underwater spectrometers, and more. NEPTUNE predates the OOI, beginning operations in 2009. It is the largest cabled ocean observatory associated with ONC and is located in the Northeast Pacific Ocean, a dynamic region that lends itself to the study of land-ocean interactions, nutrient circulations, gas hydrates, and more. In fact, the large reach of this observatory allows for the observation of a wide range of environments, from coastal waves to deep-sea hydrothermal vents. Like the OOI's Regional Cabled Array (which lies off the coast of Oregon), one of its nodes is at the Juan de Fuca Ridge, while others are fixed at the Clayoquot Slope, Barkley Canyon, the mid-plate at Cascadia Basin, and on the continental shelf at Folger Passage.[20] Peter Phibbs and Stephen

Lentz, the designers of ONC's NEPTUNE Canada, have reiterated that their efforts to wire the ocean will address a universal scientific need for continuous long time-series measurements and multidisciplinary experiments.[21]

Located a bit further south on the Juan de Fuca Plate, the OOI Regional Cabled Array (RCA) is designed to study "globally significant oceanographic processes" by tracking multiple variables in a dynamic region over a long period of time.[22] This includes studying biogeochemical cycles, fisheries, tsunamis, carbon flux, and plate tectonics. Locally, however, the Axial Seamount is the most "magmatically robust volcano" on the Juan de Fuca Ridge, which makes it a strong candidate for the study of fluid-rock interactions, geodynamics, and turbulent mixing processes.[23] Notably, the part of the RCA infrastructure at the Axial Seamount is able to provide live detection of seismic data, and has been called "the most advanced underwater volcanic observatory in the world ocean."[24]

Moving into the Pacific, the ALOHA Cabled Observatory (ACO) was developed and run by University of Hawai'i at Mānoa in order to observe the abyssal environment north of the islands, a spot that is

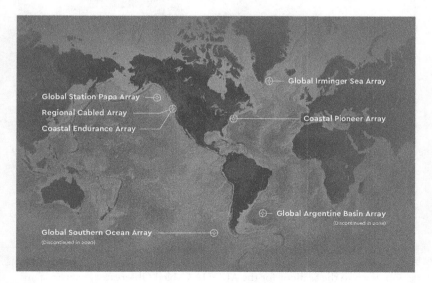

Figure 15. A 2020 map of the Ocean Observatories Initiative cabled arrays. Image by Ocean Observatories Initiative.

considered representative of most of the world ocean (see Figure 16). In particular, it has been used to study carbon cycling and biogeochemical transformations in the context of climate change. In addition, temperature measurements from the site have contributed to the charting of repeated cold events. Operational since 2012, the ACO is currently the deepest operating ocean observatory in the world, at 4,728 meters, and is known for its live-streamed hydrophone data, among other sensors and instruments.[25] Every seafloor observatory includes a primary infrastructure consisting of fiber-optic cables, shore stations, junction boxes,

Figure 16. Marine technician Drew Cole prepares to load and launch an "elevator" that will take new instruments (including a camera, CTD, and cable spool) down to the seafloor for the ALOHA Cabled Observatory as part of an ACO expedition in May–June 2021. Photograph by Meagan Putts with the ACO group.

and observatory modules, as well as secondary infrastructure, which includes the components between user ports and specific sensor instruments. The design and deployment of such delicate technologies is no easy task, and the resilience of observatory systems is increased through practices that take advantage of existing infrastructure. In fact, seafloor observation networks are largely possible in the first place because the global internet itself exists partially underwater. Nicole Starosielski charts this network of fiber-optic cables in her volume *The Undersea Network*, noting the ways in which submarine internet infrastructure is made invisible to the public.[26] Many previous generations of cable technology are now defunct, leaving miles of obsolete cabling at the ocean floor.[27] This undersea network largely provides the basis for subsequent cabled observation projects.

For example, the ACO was built on the backbone of a decommissioned first-generation telecommunications cable terminating on Oahu. The HAW-4 cable, originally owned by AT&T, had been working for twenty years prior to its repurposing.[28] The existence of an old transoceanic cable system provided a cost-effective means of providing power and communications bandwidth to the observatory: "Since the cable is already in-place and is designed to operate for well beyond its commercial lifetime, costs of conversion to scientific use are substantially lower than for new systems."[29] Proposals for an "Oceans 2.0" by organizations like ONC likewise recommend refitting telecommunications cable infrastructure for the purposes of underwater observation. Old or decommissioned cables are also sometimes recycled by research institutions to provide support for ocean observatories. As Paul Edwards puts it, "Infrastructure is sunk into, inside of, other structures, social arrangements, and technologies."[30] Existing telecommunications infrastructure provides the template for integrated ocean observation infrastructure, and it is common to find regular collaborations between marine scientists and telecommunications workers.

This infrastructural repurposing also exists for the OOI observatories, although to a different degree. The RCA is distinct in that its primary infrastructure was mostly built and designed from the ground up. In 2014 a company called L3 Maripro manufactured and installed the primary infrastructure and secondary infrastructure for the RCA based on designs by an external telecommunications committee.[31] The primary nodes convert and distribute power and communication from the shore station to a set of junction boxes, which then extend power to a secondary infrastructure that provides access to seven observational nodes equipped with low-voltage instruments along the southern Juan de Fuca Plate. Sensors include seismometers, hydrophones, pressure devices, a high-definition video camera, a long-duration fluid sampler, a mass spectrometer, benthic flow meters, and more.

The array was then also piggybacked on top of an abandoned shore station in Pacific City, once owned by a telecommunications company that had gone bankrupt. Additionally, the site of the Oregon cable was itself selected based on the presence of previous infrastructure and "its proximity to the historic Newport Hydrographic Line that has been sampled regularly since 1961."[32] According to Deborah Kelley, director of the RCA, the selection of OOI sites hinged heavily on not only scientific merit but also a calculation around costs and potential conflicts and complications, determined with the help of Navy engineers, telecommunications experts, and fishermen.[33] Running from the main shore station of the RCA are two backbone cables: one to the Axial Seamount, and another to the base of the Cascadea subduction zone and then offshore of Newport, Oregon. The directors note, "The backbone cable is comprised of approximately 900 km of telecom industry subsea electro-optical cable that provides 8 kW of power and redundant 10 Gbps data communications to each primary node."[34] This does not reflect the most current fiber-optic cable capabilities. The standard bandwidth for submarine telecommunications cables in the 1990s was 10 Gbps, but by 2010 the telecommunications industry itself was regularly using fibers that could transmit ten times that amount.[35]

While it is a cost-effective means to build ocean networks, the use of recycled cables may change as submarine cable manufacturers adapt their telecommunications systems for scientific use. There are already proposals for Scientific Monitoring and Reliable Telecommunications (SMART) cables, which seek to add temperature and pressure sensors and other kinds of instrumentation at regular intervals onto commercial submarine telecommunications cable systems.[36] These proposals would attach sensors to the cable itself. More than merely transmitters of signals, undersea fiber-optic cables, with the help of instrument attachments, could multitask as both transmission devices and sensing agents.[37]

This technical trajectory mirrors that of media infrastructural systems at large, whose functioning is largely and purposefully invisible to the public. We tend to ignore the presence of instruments, wires, cables, control stations, and infrastructural labor in favor of the data that the infrastructures collect and disseminate. But how do the stories that we tell ourselves about the deep seabed interface with the nitty-gritty of physical, infrastructural practices? As I will show next, this pursuit of remote control over aquatic environments has a tendency to center a teleological notion of technical humanity.

Homo aquaticus 2.0 and the View from Below

Designs and justifications for cabled observatories lean on both senses of the observatory as a networked entity and charismatic in situ observer. Both of these novel underwater capabilities have once more made the idea of an amphibian humanity irresistible, contributing to perceptions of cabled observatories as transformative infrastructures for the human species. Accompanying these futuristic imaginaries is a consideration of how ocean infrastructures presently act as zones of contact between humans and marine life, technology and water, and local and global stakeholders.

Evolutionary theories about the aquatic man have circulated for decades. The now-discredited aquatic ape theory, for instance, gained traction in the 1970s and 1980s as writers, zoologists, and anthropologists like Alister Hardy and Elaine Morgan speculated about possible relationships between distinct hominid features (hairless bodies, subcutaneous fat) and their perceived utility for aquatic existence.[38] Speculations about aquatic man apply evolutionary perspectives to the future and to technology as much as they do the past, however. In her recent history of *Homo aquaticus*, Helen Rozwadowski notes that evolutionary perspectives naturalized the idea of employing the underwater as a space for productive ends as well as for survival in the face of environmental crisis: "*Homo aquaticus* actively promoted novel cultural views of the ocean, initially as a limitless frontier, but shifting in the 1970s to a place of last resort for human survival."[39] Indeed, for a long time the notion of a *Homo aquaticus* was tied to Western colonialism in new resource frontiers.[40] DeLoughrey mentions two perspectives: *aqua extractio*, which "largely figured the ocean and its resources as subject to the exploitation of discrete national territories"; and *aqua homo*, a cultural approach to the ocean as a masculine, "historical space of transnational capital, empire, and slavery."[41] As I discussed in the previous chapters, a prioritization of material extractions in the design and implementation of undersea media technology has perpetuated both *aqua extractio* and *aqua homo* perspectives into the present day, working alongside visions of survivalist infrastructure.

Cabled seafloor observatories fundamentally change the terms by which we are able to occupy, exploit, and experience the seas. This includes both spatial and temporal modes of experience. First, the building of the global seafloor observation network is frequently justified to the public through the charisma of in situ ocean observation. Localized views from below simulate a feeling of being immersed in water, extending volumetric views from the skies to the seas. Musing on a trip to an aquarium, Eva Hayward describes the somatic experience of immersion as producing a sense of cohabitation, as opposed to mere representation.[42] Like the aquarium, seafloor observatories can also produce a sense of cohabitation by mobilizing in situ views and emphasizing live telepresence.[43] There is an inherent entertainment value as well as scientific benefit to immersive seafloor cabled observation. For instance, video monitoring from the ALOHA observatory has led to the discovery of new benthic species as well as animal behaviors never before recorded. As investigator Bruce Howe explains, six weeks of video monitoring led to the identification of fifteen species in addition to a sighting of a "deepsea lizard fish attacking an aristeid shrimp."[44]

Human desires for glimpses into deep-sea action like this are primed through nature documentaries such as *Blue Planet*, which frequently play up the drama of predator and prey interactions. Underwater film historian Margaret Cohen describes these narrative tropes as "liquid fantasies" that blur the divide between documentary and fantasy. Writing about *Blue Planet II*, she explains, "We find our fantasies about the sea, in large measure created by film and television, realized in live action."[45] As a result, documentary-style "infotainment" has become a staple format for fostering ocean education and empathy and has come to reinforce certain kinds of aesthetics and narrative expectations around ocean images. This includes reinscribing Western desires to capture and consume pristine, exotic images of ocean space. LED lights flood the dark depths to reveal luminous jellies, vibrant hydrothermal vents, fluorescent squids—colorful worlds that come into being only as we reveal them with a blinding, unnatural light. Using light, in situ images "enrich" human imaginaries of the ocean, even as light and sound pollution may be experienced by deep-sea critters as an ecological "fouling" agent.

Our perception of the deep sea as either fertile or deserted entirely depends on this mediation. These illuminated views are privileged both in popular science communication efforts and in popular science fiction. Just think, for instance, of the luminescent aquatic worlds in the recent Hollywood film sequels *Avatar: The Way of Water* (2022) and *Black Panther: Wakanda Forever* (2022), which both imagine "blue" underwater civilizations peopled with technologically advanced (and very blue) humanoids. These films return us to an indigenized version of *Homo aquaticus* (although it is undoubtedly a Western fantasy of an indigenized subject). Interestingly enough, both film series include narratives driven by extraction, and in *Wakanda Forever*, deep-sea mining is the main event, which perhaps speaks to an increased public awareness of seabed mineral extraction. Disappointingly, the movie does little to address or remedy this fictionalized version of the problem, despite placing it as a central motivation for the plot. In more ways than one, fantasies of aquatic humanity cross between fiction and extant scientific work in our oceans, but with important differences.

Unlike conventional ocean films and documentaries, the live and continuous operations of cabled observatories also endow them with a distinct power as truth tellers and witnesses. Gabrys calls live feeds or camera-as-sensor assemblages "imagers," a form of informatic imaging that functions within distributed networks of capture.[46] Early in situ live feeds such as critter cams or even the BP oil Spillcam were imagers that had a significant impact on public opinion about offshore oil. Nadia Bozak writes, "The Spillcam was instrumental in galvanizing media, scientists, and citizens against the criminal oil giant caught on its own surveillance video red-handed, in real time. . . . The Spillcam 'caught' more than just BP in transgression; it caught the hydrocarbon world."[47] The incidental capture of human events such as oil spills exemplifies the ways in which imagers are often thought of as passive, unfiltered, and objective witnesses. As such, imagers are capable of capturing not only nonhuman secrets but also the foul secrets of marine industry.

This function of in situ imagers as potential witnesses to accidents, crimes, and incidents of environmental fouling is latent in every cabled observatory. For instance, James Potemra, one of the founding principal investigators for station ALOHA, told me that documentarians have been interested in station ALOHA because of its potential to detect events like the Malaysia Airlines Flight 370 disappearance in 2014.[48] This idea did not come from any prior case of observatories acting forensically; it was pure speculation. If we had an underwater network in place in the Southern Indian Ocean at the time, perhaps we *would have* heard it crash. Military seafloor observatories like China's ambitious "Underwater Great Wall," composed largely of cabled acoustic sensors at the bottom of the South China Sea, operate through the same premise.[49] Like a spider's web, a seafloor observatory network doubles as a surveillance system.

These speculations attest to the fact that the temporal capabilities of cabled observation are seen as more revolutionary than its mere spatial occupation of the seafloor. By making possible continuous virtual presence of human technologies in the ocean on a global scale, seafloor observatories create an entirely new world of ocean questions and potential answers. For example, scientists studying a typhoon season can now directly observe precipitation and temperature changes over the course of a storm, as well as compare storm data over several years, enabling better modeling and prediction. The excitement over this potential transformation in knowledge is reflected in the descriptions for pioneer observatories. Take, for instance, this description from ONC: "Smart Ocean™ Systems are a paradigm shift in how science and ocean monitoring is conducted. They address the limitation of conventional technologies to allow continuous year-round, sub-second observations with dozens of measurement types, accessible through the Internet to any audience."[50] A 2001 paper by H. Lawrence Clark (NSF) announcing the OOI seafloor observatory networks states it more simply: "A new system of observatories, accessible to all investigators, would facilitate the 'temporal' exploration of our oceans."[51] What makes cabled ocean observation "smart" therefore has much to do with its long-term observation and real time data.[52] These observatory proponents seem to turn away from imaginaries of spatial colonization, implying a transition of knowledge from ex situ (off-site experimentation) to in situ (on-site research), from spatial expeditions outward to temporal explorations within.

Certainly, space and time are always in inextricable relation. Orit Halpern, Robert Mitchell, and Bernard Dionysius Geoghegan explain that "smartness colonizes space through the management of time."[53] In another example, the developers of the ACO evoke both the spatial and temporal innovations of the observatory: "As new cables continue to be laid between continents, providing the fabric of interconnectivity required by our rapidly evolving technical society, an entirely new paradigm of ocean and geophysical measurements may be enabled."[54] Sustained in situ capabilities mean that the wired abyss is not simply a tool but a transformative process with new capacities and modes of sense, suitable to the ambitions of a technical terrestrial society. Beyond mere knowledge gains, however, cabled observation entails shifts in deep-ocean imaginaries at large, and diverge in important ways from older, visual mediations of ocean space, which tend to highlight interspecies encounters and maintain a greater sense of the deep sea's unique materiality and milieu.[55]

Narratives of underwater techno-evolution and techno-utopianism were given momentum by proposals from oceanographers and ocean engineers that stressed the novelty of a view from the bottom. In the past, ocean knowledge was largely driven by sampling, as sensors and other instruments were limited by battery life and by the limitations of survey ship time. In fact, while data continuity has always been the goal, much of what we know about the deep seabed is still derived from samples—mere pieces of a whole. Continuously powered instruments, by contrast, create enormous volumes of data that allow scientists to pick the scale at which they sample information and to store data for an unlimited amount of time for future studies. Once, humans required "passports" to the sea, and technologies like diving suits and deep submergence vehicles that created what Melody Jue calls "conditional amphibiousness." Today, ocean observatories promise long-term presence in water, enabled by internet technology.[56] It could therefore be said that cabled observatories participate in an *unconditional amphibiousness* that seemingly goes beyond visitation. Arguably, the desire to construe smart technology as continuous regardless of physical constraint simultaneously defines the internet as an abstract, even placeless technology—one that can extend anywhere, anytime.[57]

Like putting a man on the moon, it is easy to imagine unconditional amphibiousness as a turning point in human history. An influential text by oceanographers Paolo Favali, Laura Beranzoli, and Angelo De Santis on the subject of seafloor observatories asserts, "The development of our understanding of the ocean has been hampered by our terrestrial existence as a species. . . . However, a continuous interactive presence in the ocean, more analogous to how our knowledge and intuition about terrestrial environments has built up, has been elusive."[58] The statement suggests that technological permanence in the deep sea equates to a more-than-terrestrial existence for the human species. Technologies are, for this group of stakeholders, simultaneously meaningful as extensions of humanity and as extensions of terrestrial knowledge. The effusive rhetoric around cabled observatories begs the question: How different or new, really, is this utopian vision of a world for *Homo aquaticus* from our ways of knowing on land?

At its core, the desire to see ocean observation systems as revolutionary and groundbreaking projects capable of transforming human knowledge speaks to the urgency of our global ocean questions. As the ocean-observing community increasingly acknowledges humanity's role as ocean polluters and fouling agents, observatories give scientists the ability to make more minute kinds of observations that lead to better hindcasts and forecasts of internal ocean processes—things that shed light on climate change. The ability to form knowledge about multiple temporal scales and the circulation of sediment, nutrients, carbon, methane, and heat thus has a bearing on our own environmental futures. Cabled ocean observatories transform how we experience large-scale oceanic events, shifting from a perspective of catastrophe and rupture to one of continuity.

However, by transforming the oceans into an ocean of data, cabled observatories also perpetuate terrestrial cultures of dataism as integral to underwater humanity. Like aquariums and other underwater media interfaces that "turn the world into information, not experience," this data interface has the side effect of constraining the possibilities for ocean viewership. As Ann Elias puts it, "it produces a viewer disposition to look

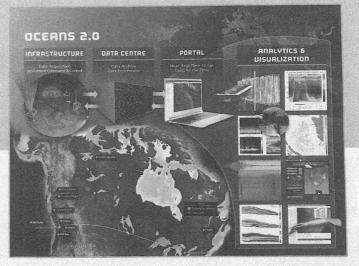

OCEANS 2.0:

AN INTERNET OF THINGS FOR THE OCEAN

You've probably heard of the "Internet of Things": a connected network of everyday objects that talk to each other, such as cars, kitchen appliances and heart monitors. But did you know that the Internet of Things also extends deep underwater off Canada's three coasts?

OCEANS 2.0

| INFRASTRUCTURE | DATA CENTRE | PORTAL | ANALYTICS & VISUALIZATION |

Think of it as a Fitbit for the ocean. Made possible by world-leading Oceans 2.0 data management software, Ocean Networks Canada's (ONC) infrastructure is continuously monitoring the pulse and vital signs of our deep sea and coastal environments. Thousands of Internet-connected sensors gather real-time continuous data—everything from temperature, salinity, tides, seismic activity to underwater noise levels and video footage.

Gathering 250 GB per day from an expanding network of Internet-connected instruments, Oceans 2.0 mak ocean data available to scientists, communities, a leaders, helping them to make informed decisions abi climate change, earthquake and tsunami detecti marine safety, life in the ocean, and more. Ocea 2.0 operates like an Amazon shopping cart, except free. Browse through the data from ONC's hundre of underwater and land-based sensors, select information you want, and then confirm your orc which is downloaded to your computer.

Figure 17. Ocean Networks Canada uses the phrase "Fitbit for the ocean" on its website and promotional materials to describe cabled ocean observatories, as can be seen in this description from one of its pamphlets. Photograph by author.

and record but not necessarily to develop empathy."[59] Although there is undoubtedly a two-way relationship of encounter between humans and marine species inherent in the building of this infrastructure, datalogical mediations tend to filter out nonhuman perspectives, reinforcing ideas of human agency and exploration in alien space. Given the data extractivism inherent in this technological regime, it would appear that *Homo aquaticus* 2.0 is paradoxically enabled through *aqua homo*, and to a large extent, a reading of the ocean in terrestrial terms.

As discussed in previous chapters, medical analogies for ocean observation further position the ocean as an ailing patient, oceanographers as benevolent doctors, and ocean technology as monitors of health. Seafloor observation networks from the deep sea to the coasts are no exception, as they are now being described as "ocean Fitbits," or stethoscopes that "take the pulse and vital signs" of the global ocean (see Figure 17).[60] Unlike the pacemaker discussed in chapter 4, however, the Fitbit is a metaphorical choice specific to a culture premised on data extraction and capitalist enclosures in digital space. The ocean's new Fitbits help transform the sea into an anthropomorphized image of a fragile, yet fertile being. While marine life is understood as always already disappearing in the Anthropocene, it is further framed by conservation efforts as an incubating potential that can be nurtured and, with the aid of the right technology feeds, eventually grown into a bioenriched landscape. Despite all the fanfare around humanity's underwater evolution, it seems the effect of these observatories on conservation philosophies do just as much or more to make the oceans like us, rather than to make us into aquatic beings. This dynamic becomes more complex, however, when we shift our perspective away from technophilic speculation and toward the contestations around technological access in addition to the materiality of ocean observation technologies themselves.

Making the Deepwater Feed FAIR

As is evident in the multiple layers of mediation they offer, from live camera feeds to internet-connected databases, cabled observatories carry a unique set of affordances and limitations when it comes to expanding humanity's access to and understanding of deep-ocean environments. However, despite lofty language around humanity's existence as a species, or fantasies of unlimited public access to global data, the high costs of building deep-sea infrastructure create unequal participation among human actors.

The questions posed by the deep sea are often planetary in nature, and the negotiations that determine the future of the seafloor happen locally, regionally, and internationally. In contrast to visual epistemologies that celebrate the power to unveil, sensory data streams are not self-evident but must be filtered and translated by specific human actors and institutions. As a result, one cannot speak of the benefits of cabled observatories without acknowledging the power dynamics at play in this space. History warns us that the utopian ideal of *Homo aquaticus* might not be a truly inclusive one. In the past, terrestrial observatories served as imperial projects that devalued other ways of living with and identifying with land and sea. This existing dynamic is compounded by the ties between oceanography and extractive industry. Just as was the case for reflection seismology, modeling efforts, and AUVs, there are close connections between seafloor cabled observation and the expansion of seabed extractive infrastructures.

As a response to what is perceived as an underdeveloped deep sea, the concept of a seafloor observatory has its roots in nineteenth-century observatory science, beginning with astronomical observatories. Astronomical observatories were mostly optical technologies that helped to reconfigure and objectivize vision during this time. This mode of representation had widespread implications. Nineteenth-century observatories brought to prominence particular techniques that took on social and cultural meaning, defining what it meant to participate in a modern world-system: precision measurements, expensive endowments, and perhaps most importantly, "the construction and maintenance of extensive networks in which observatories were key nodes."[61] Scientific observatories were progenitors of the very idea of a network, a mode in which the observer "looks with his own eyes, but sees with the eyes of the collective."[62]

The technological sublime of the observatory has also been premised on the neocolonial expropriation of indigenous lands. David Aubin, Charlotte Bigg, and H. Otto Sibum argue that observatories are spaces of scientific practice that act as pillars of the state: "It is hardly worth insisting that observatory techniques—and not just the techniques of medicine and natural history that are often studied by historians of colonial science—also proved indispensable in the constitution of large overseas empires."[63] As the authors note, observatory sciences such as astronomy, geodesy, hydrography, and meteorology in the nineteenth century were highly implicated in settler activities and used for the purposes of imperialist conquest and colonial administration. This association between power and scientific networks has not diminished with time, as struggles over the construction of modern observatories both above and below water continue to define and impose a Western, hegemonic ideal of technological modernity worldwide.

Notably, there has been a robust indigenous critique of observatories in Hawai'i, which also happens to be the site of two major ocean observatories, including the ACO. Tensions between indigenous Kānaka Maoli and astronomers came to a head in 2019 with the month-long protests over the Thirty Meter Telescope, a proposed, $1.4 billion observatory on Mauna Kea peak. This footprint of observatories and telescopes encapsulates the settler-colonial relations between Western scientists and local communities that rely on their lands and waters for everyday resources. Nonviolent resistance to the Thirty Meter Telescope has provided a template and countered narratives that normalize projects like observatories as necessary infrastructures of the future. These protests put into question, for instance, the tendency to equate networked observation with democratic participation in information technology.

Today, discourses of global cooperation, commoning, and shared research continue to be front and center within the ocean-observing community. This call is reflected in the FAIR Data Principles, a general (not ocean-specific) scientific guideline that emphasizes findability, accessibility, interoperability, and reusability.[64] The main principles of FAIR are also related to other practical approaches to ocean infrastructure that include the "big science" argument, or the idea that "no single scientist or group of scientists should be given unrestricted, private access to research infrastructure of that scale. . . . Its use should be shared, optimized and audited."[65] However, this perspective has not always been mainstream. Kelley explains, "Growing up, you hoarded your data because that's how you made your name. Now, the evolution is data for all."[66] Helmed by researchers like Kelley, the first generation of cabled seafloor observatories now internalize mandates of access and sharing.

At OceanObs'19, Ambassador Peter Thomson, UN Special Envoy for the Ocean (responsible for leading UN advocacy and outreach around the sustainable use and conservation of the ocean's resources), spoke of a "fully integrated ocean observing system" freely available for the "common benefit of the people on the planet." OOI also uses this rhetoric: its system "provides 24/7 connectivity to deliver ocean observing data to anyone with an Internet connection free of charge."[67]

In the best-case scenario envisioned by scientists, adhering to big science and creating a data commons would facilitate scalar connections and interoperability between existing networks. The three North American prototypes that I mentioned earlier are all local or regional systems that imagine possibilities for just this kind of connection. Individual networks may work together within regional networks, just as the OOI RCA does with ONC's NEPTUNE and VENUS arrays, to track meso-scale events like anomalous temperature rises, El Niño and La Niña patterns, or harmful algal blooms. Similarly, anticipating future transnational data collaborations, instruments from observatories like NEPTUNE Canada can be connected to the ACO user ports, located in the deep sea off the coast of Oahu. However, as many indigenous observers note, the global potentiality of networked infrastructure is often used problematically to signify technological modernity as inherently enriching, and to dismiss existing observations on a local scale.

At OceanObs'19 there was significant pushback against the idea of a global smart ocean from indigenous delegates, as well as from Dr. Juliet Hermes of the South African Environmental Observation Network. Traditional ocean knowledges include many ways of knowing and are often highly relational, place-based, and collaborative. Many of these epistemologies, however, do not fit in easily with the premise of cabled ocean observation networks. Indigenous methods of monitoring can also be more qualitative, and include knowledges of species migration, travel routes, cyclical weather patterns, and unusual events. In addition, sensing is embodied, and includes touch, feel, and sight.[68] Unlike most Westerners, Kānaka Maoli also see land and sea as continuous spaces. This worldview is apparent in the traditional Hawaiian Aha Moku system of sustainable resource management, wherein water resources and land resources are treated together.[69] Paulokaleioku Timmy Bailey, an indigenous speaker at OceanObs'19, explained: "If you are to take a line and

draw it from the heavens to the oceans, you'll see that there are com-
monalities between bird and fish species. What is on land is duplicated
in the oceans."[70] Perhaps it is because of this connected perspective on
the environment that in situ ocean images do not define ocean kinships
for indigenous Pacific Islanders as they do for Western audiences. This
view holds that not every good observation system needs continuous
data collection or internet-enabled networks, and science is not the only
mode of observation that can produce documentation.

Shelley Denny, a Mi'kmaw ocean researcher, remarked that in the
ocean-observing community, the term "integration" is often used as a
synonym for "assimilation."[71] I could not help but think back to smartness itself as a neocolonial worldview where "any change can be techni-
cally managed and assimilated while maintaining the ongoing survival
of the system."[72] By contrast, Denny's talk emphasized that incorporating indigenous voices should not be about assimilation but rather about
listening and "two-eyed seeing." Some audience members even made
the point that mediation itself conditions the possibility for extraction:
we do not exploit the places that we do not know about. This is a very
difficult truth to acknowledge for scientists who are trained to pursue
objectivity through the accumulation of more and more data. A course
must be found in between these two extremes.

Researchers like Denny, while celebrated in the 2019 iteration of
the conference, have been historically excluded from the global ocean-
observing community. Now, they have emerged to "show a community
that thinks it knows best that it does not."[73] One of the most important
lessons from indigenous ocean observers might be a critique of the fantasy
of a techno-aquatic humanity at the heart of the push for cabled ocean
observation. For Denny, Bailey, and others, an amphibious humanity
does not require unconditionally amphibious technologies; we are all
already connected to our oceans. For instance, Joshua Schuster writes of
the "extensive and intimate connection to coral" held by native communities in coastal areas. Ocean inhabitants are centered in Native Hawaiian cosmologies, which foster relations with organisms like coral in both
spiritual and practical terms.[74] Centering technological amphibiousness
has the effect of marginalizing these existing possibilities for an offshore
or amphibious humanity.

At the outset, the friction between Western and indigenous scientists seems to reinscribe an uneven power dynamic between local and global observation. Resource frontiers are not a mere mishmash of local and global sensibilities enabled by cross-cultural collaboration. Rather, collaboration and the broader aspirations of global connection and discovery can end up eroding or covering up the local.[75] As Anna Tsing explains it, this displacement or deterritorialization of the local is often experienced as a detachment of cultural practices of place and a loss of the natural. Indeed, it strikes me that what Denny expressed is a warning against this kind of erosion. As techniques of global modernity, ocean networks imply a change of perceptual technique, which can facilitate an experience of staying still.[76] To account for this loss of a sense of place, environmental humanist Ursula Heise advocates for a sense of planet, a shift toward "eco-cosmopolitanism" that transcends the territorial.[77] Though Heise does not articulate her argument in terms of indigenous perspectives, a sense of planet is also seen as key to a decolonial approach to environmental management. However, challenging assumptions that global modernity leads to irretrievable losses of the local, I would contend that one does not need to bury the local in order to see this planetary connection.

The dilemmas that scholars like Heise and Tsing pose around globalization and deterritorialization already suggest that a binary perspective that reduces the local and global as antithetical to one another is overly reductive. After all, scale is, as Tsing argues, a non-neutral cultural frame that is brought into being through divergent and contested claims.[78] While globalism has often been used as a rhetorical means to compel participation into Western scientific networks and marginalize the scientific contributions of native peoples, indigenous critiques of Western science do not amount to a wholesale rejection of globalism or of networks. Far from it—even while they study the dispossessions of global modernity, indigenous scholars, for instance, insist that indigeneity is both local and global.[79] We can take a lesson here from the Kānaka Maoli, who see nature not as merely a local manifestation but as an extension of the self. A decolonial perspective might aim to maintain environmental relation through every scale and every geography; movement across space becomes an act of both self-extension and a knowledge practice. This implies the ability for coastal or local knowledge to transgress the colonial, casting it as a multifaceted relationship to nature and to a global world.

Marisa Duarte argues in her book *Network Sovereignty* that large-scale digital networks including social media have been key to fostering a recognition of indigenous epistemologies and ways of being. She explains that countering colonial governance requires "understanding when, where, and how autonomous Indigenous peoples can leverage information flows across ICTs for the purpose of meeting social and political goals, in spite of the forces of colonization."[80] Emergent networks of ocean observation must be designed to empower, rather than work against, indigenous epistemologies and ways of life. If we acknowledge that cabled observatories are not just global ambassadors for technoscience but also tools for telling stories at multiple scales, then we can see the ways in which scientific storywork might operate in tandem with a wide variety of storying forms.[81] In the next section I turn to maintenance and repair as an area in which we can see a more expansive scientific approach to infrastructural storywork.

From Homo aquaticus to Biofouled Humanity

To keep things running, cabled seafloor observatories are equipped with alerts and alarms, backup generators, and other monitoring technologies.[82] OOI researchers explain, "System health and status for cabled infrastructure is closely monitored, 24/7 in real time from shore. . . . Across the facility, instruments and platforms are monitored for safety, functionality, and basic data quality."[83] The health of the observatories, like that of the ocean, must also be continuously monitored and optimized. Maintenance activities are an important site from which we can analyze relationships between *Homo aquaticus* and *aqua homo*, as this is where researchers directly contend with the frictive interface between human technology and ocean agencies.

Every smart system requires regular maintenance, but when it comes to aquatic observatories, a maintenance cruise entails enormous efforts in organization and planning. When breakdowns happen in benthic environments, they cannot be fixed immediately. Potemra spoke about the stakes of ensuring that instrumentation works in the initial stages: "If things fail, they fail immediately because of the pressure. If things are working for several days, they keep running for years." Typically, maintenance cruises happen only once a year, and, as with other kinds of oceanographic research, the cruise schedule is limited and cost prohibitive.[84] University-National Oceanographic Laboratory System (UNOLS) vessels—large ships owned by the Navy, NSF, and other oceanographic institutions—sail annually to maintain observatories like the OOI RCA. OOI researchers explain, "The degradation of mooring components, biofouling of instruments, and depletion of batteries on the uncabled profiler moorings are the main drivers of the OOI moored array maintenance schedule."[85] Often, maintenance cruises carry ROVs to help accomplish tasks at the seafloor. In addition, the shore stations occasionally encounter issues due to large storms.

Maintenance expeditions are some of the most well-documented, visible aspects of oceanographic research. Cruises are multifunctioning, acting as opportunities to fix infrastructure, take measurements, gather data for novel research questions, and document oceanographic activities for the public. ONC and OOI both use Instagram and their websites to publicize their maintenance cruises and turn them into educational opportunities (see Figure 18). The ACO also documented its 2011–2015 deployment cruises on Instagram, although it has not been updated since. Like other ocean researchers, cabled observatory maintenance workers are accorded status as intrepid explorers and technology experts who must battle the challenges of weather, currents, and low visibility. This kind of spectacular maintenance serves in large part to demonstrate the biodiversity of the deep sea and foster empathy for ocean ecosystems, while simultaneously promoting underwater connectivity as a means to protect those same environments. ONC posts appear on Instagram every two or three days and offer detailed descriptions of instrument deployments using hashtags such as #wiringtheabyss or #knowtheocean. Posts most often contain images of workers and equipment, including in situ images of cabled infrastructure. These are mixed in with images of octopuses, anglerfishes, ocean perches, squids, whalefish, and other organisms. Like other ocean outreach programs, maintenance cruises provide an opportunity for oceanographers to humanize their labor, as well as introduce the general populace to deep-sea ecosystems. ONC, for instance, has a "Critters of the Abyss" spotlight featuring curiosities encountered during maintenance and cable-laying expeditions.

Figure 18. The OOI regularly documents its maintenance and repair activities over its Instagram and website. This image of the Irminger 8 deployment emphasizes the teamwork involved. Image supported by the National Science Foundation's Ocean Observatories Initiative, a major facility fully funded by the NSF. Photograph by Dee Emrich at WHOI.

In an April 9, 2022, post about the Remotely Operated Vehicle for Ocean Sciences (ROPOS) deck in the Cascadia Basin, a helmet jellyfish appears in the foreground of an image of a gridded platform, 2,600 meters underwater. In a post from December 17, 2019, a deepsea skate slides by an ONC frame at the Endeavour segment of the Juan de Fuca mid-ocean ridge. Spider crabs are also often caught playfully climbing on the jungle gyms of deep-ocean equipment. In addition to explaining the function of these infrastructural components, the posts delve into the biology or ecology of the organisms captured by the camera.

Even more spectacular than mobile creatures like glass octopuses, ratfish, or spider crabs is the appearance of biofouled infrastructure (see Figure 19). Biofouling is a key concern for maintenance because these accumulations can crowd the space of a camera lens or weigh down instrumentation and interfere with a sensor. For long-term underwater infrastructures, antifouling, or the removal of biofouled buildup, is frequently featured in maintenance activities as a form of technological hygiene. My attention to biofouling is not merely proclamation of ocean technology's naturecultural existence, however; biofouling and bioenrichment are values that have resonated differently at particular historical junctures. For much of Western history, mariners were even less generous toward biofoulers, and this contentious relation drove hundreds of antifouling paint patents and sheathing solutions during the height of European colonialism and British naval dominance in particular.

While the "foulness" of biofouling indicates the historical tendency to perceive marine life as a problematic barrier to ocean observation, recent popular mediations are much more ambivalent about the phenomenon, verging closer to a friendly antagonism. This is suggestive of the way different mediations of naturecultural hybridity—from narrative filmmaking to maintenance documentation—enable distinct modes of environmental perception and action. In one ONC post from December 2020 (see Figure 20), ONC scientist Steve Mihaly peaks out from behind a colorful smorgasbord of marine life clinging to and obfuscating

Figure 19. This post from the OOI image gallery highlights the biofouling on the Inshore Surface Mooring over the course of a year at the Pioneer Array. Image supported by the National Science Foundation's Ocean Observatories Initiative, a major facility fully funded by the NSF. Photograph by Dee Emrich at WHOI.

the very shape of the platform itself. The caption includes a quote from Mihaly: "Even on deep sea dives I haven't seen anything to compare. You could touch the animals, soft, furry, very hard, sharp, fast, slow. . . . It was dizzying."[86] The frequent expressions of reverence for the diversity and beauty of marine biofoulers on these pages is sharply incongruent with the term "biofoul" itself. One commenter even posed a revision on the term *biofoul*: "bio-enrichment not fouling." These texts reveal the potential of ocean media to foster respect, admiration, and knowledge of ocean ecosystems, even when they interfere with human activity. Together, the animal and the technological components of the image seem to constitute an entity that is more than the sum of its parts.

Thick, textured masses of marine biofoulers also evade normative perceptions of marine life as aesthetically beautiful, fragile, or inquisitive. Instead, colonial organisms are simply spectacular for their strangeness. What this post captures, ultimately, is the sheer excitement of a novel multispecies encounter. Biofoulers themselves become part of underwater infrastructure, as they form kinships with the maintenance workers who venture into this poorly understood space. This relation of mutual fouling and mutual enrichment troubles tendencies to think of infrastructure

Figure 20. Instagram post from @Ocean_Networks depicting biofouling on the Baynes Sound platform during a maintenance trip on December 9, 2020.

in terms of human agency and anthropocentrism. The work of "infra-structuring" here is more than a matter of holding things together; it is also matter attracting matter.

Rather than isolate the marine animals in their environment as popular documentaries do, these images depict a wide variety of encounters between the human technologies and ocean life and the spatial zones in which underwater media production happens. The contact zones of infrastructural maintenance complicate prevalent discourses of invasive species and invasive humans, directing us instead to a mutual grappling-with between humans, technologies, and marine species. In fact, the persistent conflation between biofouling and bioenrichment within maintenance media points to the socially constructed and symbolic maintenance of boundaries between land and sea, technology and biology, and the active negotiations that happen around these borders.

A key piece of Pratt's and Wilson's conceptualizations of the contact zone is not just its ambiguity but the way this meaning-making is shaped by asymmetrical relations of power. Indeed, it is the human crews who embed the infrastructure into ocean space, holding the power to craft the images and the narratives that accompany them on social media. The power dynamic of the contact zone is lopsided in more ways than one. In many ways, biofouled infrastructures reflect broader histories of human exploitation of the ocean, as a technological society is also one that pollutes and thus creates environmental crisis. Scholars working in environmental media studies have already shed light on the environmental costs of a high-tech society, which includes everything from toxic mining practices for rare earth minerals to the massive amounts of pollution emitted by data centers.[87] One could say that humans are the ultimate biofoulers, displacing material from one environment to another.

It is perhaps the specter of humanity's own foulness that informs the production of biofouled, humanoid villains in popular media. The second installment of the *Pirates of the Caribbean* franchise, *Pirates of the Caribbean: Dead Man's Chest* (2006), popularized an image of the barnacled pirate as a cursed man, a human-turned-monster, doomed to serve as a crewman on the *Flying Dutchman*, a ship ferrying lost souls at sea to the afterlife (see Figure 21). Biofouled humans can also be seen in the Trench monsters of *Aquaman* (2018), as well as the writhing swamp creatures in monster movies from the 1950s and 1980s such as *Swamp*

Thing (1982) and *Creature from the Black Lagoon* (1954). In popular media, biofoul often marks a nomadic subject that does not belong or disobeys and is therefore monstrous. Biofouled lives are at once abjected parts of us, inhuman, and also more-than-human in their connections to aquatic environments. If the evolutionary version of *Homo aquaticus* was born out of colonial society, utopian imaginaries of an enriched seafloor, and the enclosures of extractive capitalism, then the barnacled pirate is its foil: lowly parasitical existence, lost treasure, and uncontainable disorder. We can see traces of both figures in the biofouled contact zones of seafloor observatories.

Beyond visions of a fouled humanity or a spectacular underwater paradise, the ocean maintenance worker becomes a relatively accessible and ordinary figure for *Homo aquaticus*. The sustainability of ocean infrastructures depends not only on the technological labor of these workers but also on the labor of public communication at observatory sites, which helps to foster empathy and curiosity for a plethora of nonhuman others. Maintenance media remind us not only of the important scientific functions of ocean observatories but also of the subjectivity and agency of the visiting octopus or the playful spider crab. These, unlike the ocean Fitbit, allow for and even emphasize the continued presence of the foreign and unfamiliar. This type of scientific storywork should not be viewed as secondary to the charismatic imagery produced specifically for popular spectacle, or to the more ambitious global data-collection

Figure 21. Pirates of the Caribbean: Dead Man's Chest (2006) prominently features barnacled pirates as entities existing in a liminal state between life and death, humanity and nonhuman nature.

efforts occurring to address issues like climate change. In contrast to ocean media that either evacuate the presence of humans in order to produce the illusion of immersion into wilderness, or fixate on the deep-sea observer as hero-explorer, maintenance videos offer a vantage point from which to observe a wide spectrum of human–ocean entanglements—interspecies relations that are at once adversarial and inquisitive, harmful and enriching, banal and spectacular.

Many Ways to Feed

I have explored the intersecting imaginaries of seafloor cabled observatories and recurring ideas of the "aquatic man" in order to suggest that speculations about human underwater futurities can extend beyond evolutionary or settler imaginaries. It is tempting to think of seafloor observatories as extensions of frontierist thinking. Frontierism, as I discussed in chapter 1, fuses mandates for technological innovation, spatial expansion, and domestication, each of which pits man against nature. Already, disconnected spaces in the ocean are construed as problematic gaps in a database, while rhetoric around building the smart ocean assumes that it is both necessary and inevitable. But while we do see desires for human occupation and control of deep-sea environments, observatories also challenge humanity's perception of itself in relation to the seas. More than merely augmenting human capacities, these technologies problematize extractive or one-way relationships between humans and the ocean. They enable a mode of mutual relationality, neither positive nor negative, between humans and marine organisms. Both alternately inhabit roles as biofoulers and sources of bioenrichment.

I titled this chapter "Deepwater Feeds" because I see the feed as an apt metaphor for benthic becomings in more way than one. They are suggestive of a relationship between intake and sustenance, between information and vitality. I think of filter feeders, which splay out their sensing limbs in the hopes of catching tiny morsels of food floating in the vast emptiness. For organisms like a deep-sea coral, to feel is to eat; feeding is both sustenance and perception. I am reminded of the works of Donna Haraway and Karen Barad, both feminist STS theorists who were fascinated by the perceptual capacities of tentacled beings. Barad was particularly taken with the brittlestar, a photosensitive invertebrate and relation of the starfish: "Brittlestars do not have eyes. They are eyes. . . . The brittlestar is a living, breathing, metamorphosing optical system. For a brittlestar, being and knowing, materiality and intelligibility, substance and form entail one another."[88] Just as the brittlestar helps break the anthropocentric idea of mediation as technologically mediated knowledge, my explorations of feeding here imply that metabolic imaginaries and life-sustaining processes, human and nonhuman, are ingrained within the most technologically advanced forms of deep-sea mediation.

Ultimately, my critique of the cabled seafloor observatory networks in this chapter serves not to discard their possible interventions into ocean knowledges but rather to refuse their unifying ambitions. The existing politics around terrestrial digital networks offers a pessimistic, monocultural view of what *Homo aquaticus* 2.0 could become. Although most seafloor sensor networks concentrate agency at the interpretive end, a truly democratic, decolonial sensing system would allow for autonomy and public access at every step of the process, from infrastructural production, to methods of observation, to database construction and access. This means opening up to futurities that accept fundamental changes to existing systems, futurities that understand the deep ocean as a heterogeneous space with its own possibilities for interspecies connection that exceed our aspirations for cyberinfrastructure. And while the question is often avoided, it behooves us to ask: What would it mean to accept, with courage, a world without unlimited feeds?

CONCLUSION
Defining Ethical Ocean Mediation

At the start of this book I proposed that the international pursuit of a deep-sea futurity has been marked by a circuitous return to extraction and to notions of human control over nature. To profit from our oceans, we had to study them. We had to engage in prospecting, an act of projection, which filtered down prospects according to a prior economic and geopolitical scaffolding. I found much of this discussion of prospects in the writings of primarily U.S.-based ocean engineers—both formal and informal—demonstrating an awareness of extraction as the key ideology structuring contemporary human relationships to our oceans and especially its deep spaces.

In 1967, Alfred Keil, then the dean of engineering at MIT, published a piece in *Technology Review* advocating for the development of underwater instrumentation, transport, and extractive infrastructure. In the article, he reflects on the decade's transition from navigation at the ocean's surface to the depths:

> The problem of the Navy in keeping tabs on the world's oceans is no different from the problem of engineers attempting to use the oceans' resources: both factions must now recognize that all the water underneath the surface is at their disposal, as well as the surface itself. The challenge to ocean engineering is to produce vehicles fit to travel and work in this medium and to harvest its riches.[1]

For Keil, ocean engineering was a field dedicated to the ambition of overcoming "the natural forces of the ocean depths," building on the previous achievements of coastal development, naval architecture, and fishing technology.[2] One of the most influential ocean engineers of his time, Keil emphasized in his numerous public and institutional engagements

189

that industrial exploitation of marine resources would be contingent on both the development of new survey technology and the acquisition of knowledge about ocean environments.

This environmental knowledge included much of what has been discussed in this volume: "ocean bottom reflectivity with respect to sonar applications, noise characterizations of marine life, bearing strength of ocean bottoms for the erection of bottom installations, settlement of bottom material which has been stirred up by activities on the bottom and fouling and stress corrosion for deep-sea installations."[3] This book has explored a number of scientific and industrial techniques for viewing and sensing the deep ocean that build from this initial premise to know and to tame the seafloor. Meanwhile, many of the images, stories, and data that we produce about the seafloor arise from the lives and values that we have accumulated as terrestrial beings. Our hopes and dreams for an aquatic or amphibian humanity are imperfectly extrapolated from the present in ways that are often mismatched with the fluid and interconnected nature of the ocean and its ecosystems.

Throughout these chapters, I have made the case that looks of discovery, remembrance, and exploitation are enmeshed in one another through shared media imaginaries of the deep seabed. When we filter ocean caretaking through the lens of blue economy, we create codependents instead of recognizing interdependence, defer to heroism and charity instead of collectivity and mutual aid, advocate for resilience instead of repair, and seek out preservation instead of cultivation. Humanity's empathy for ocean dwellers dissipates in the abstraction of petroleum surveys, which direct us to think of megafauna caught in sounding pathways as noisy obstacles. Our desire for marine knowledge is reoriented toward profit when we archive the seafloor solely in terms of human histories and human treasures. Care can even be pushed to the background in telemetry, when the life of a tag and its capacities to transmit information supersede the life of the animal. Cost-benefit analysis reigns supreme in environmental impact assessments that weigh industrial toxicity in terms of contested thresholds of harm and temporally nebulous definitions of turbulence. And in the final case, our attentions to the on-the-ground experiences and knowledges of both nonhuman others and coastal communities is given tokenistic consideration when it is gilded in the language of integration, replacing inclusion and co-production with assimilation and standardization.

While ocean stakeholders and scientists readily accept the humanitarian or environmentalist affordances of popular media storytelling such as environmental documentary or artwork, there is a relative lack of self-reflection about the ideologies and political agendas that drive methods of data collection and scientific mediation, many of which reinscribe a false divide between the "inhuman" sciences and a humanistic "cultural" sphere of storytelling. Given these and other limitations of our current ocean-observing systems, anthropologist Lesley Green argues that what's needed is an "ocean regime shift," a reclamation by science of a perspective that transcends the divisions of nature and culture. "Our time is one in which neoliberal governance and market logics have actively sought to capture the authority of science," Green explains. "To sustain the seas, scholars who care for them need to reclaim the big picture."[4] Green examines the South African context for fishery and water management to suture this gap in vision, and in doing so she comes to a conclusion similar to mine: that the media tools of science, which focus on processes such as enumeration, are ill-equipped to represent environmental issues that are the outcome of complex political, social, and economic choices by a human collective. Responding to Green's critique, a critical media studies perspective is an ideal starting point for pursuing an ocean regime shift. Media studies has long considered the politics of representation not only in terms of content and messaging but also in terms of how form, platform, and mediating techniques condition culture, paying attention to both the humanist agendas behind science as well as the nonhuman agencies that structure human imaginaries of the world.

Synthesizing these insights and acknowledging the need for a more caring orientation toward ocean observation and knowledge production, I end this book by offering four recommendations for how we might approach ethical ocean mediation and an ocean regime shift: decolonize ocean observation, build offshore humanisms, attend to the contact zones of mediation, and make space for the unknown.

Decolonize Ocean Observation

As my case studies show, dominant discourses of underwater technology have tended to struggle with perceived gaps in knowledge, aiming for comprehensive and continuous data feeds rather than questioning the material impacts of this data collection or reconsidering how this data might be interpreted or used for social and political aims. Efforts to capture the ocean through audiovisual mediation are redoubled by language from engineers and oceanographers that position underwater technology as augmentations on an aquatically disabled human sensorium. There is an assumption that ocean knowledge-making requires humans to enhance themselves with technology: breathing apparatuses, navigational systems, sensors, durable housings, and networked systems. Yet, while it is often couched in self-congratulatory lingo, the desire to fill gaps in knowledge with technological solutions is one that is rooted in ableist, technologically determinist, and Eurocentric frameworks for knowledge and credibility.

The occlusion of the humanities and of indigenous perspectives from mainstream spaces of ocean mediation and representation is not a mere oversight (as it is often framed), but is inevitably political in nature. Thinkers like Michel Foucault and Stuart Hall have emphasized the ways in which representation and the production of knowledge are rooted in the operations of power. Writing on Orientalism, Hall explains, "Power not only constrains and prevents: it is also productive. It produces new discourses, new kinds of knowledge (i.e., Orientalism), new objects of knowledge (the Orient), it shapes new practices (colonization) and institutions (colonial government). . . . Power circulates."[5] Indeed, one could argue that there is an interspecies Orientalism that operates in the deep sea, where marine inhabitants are prone to Western exoticization. The power dynamics around our representations of the deep ocean are particularly tangible in representations of difference—of others like our oceanic kin—as alternately monsters or as helpless victims in need of saving. Meanwhile, widely used terms like "blue economy," a hegemonic approach toward sustainability, are rooted in this circulation of power, the Western production of deep-sea knowledge, and the advancement of particular institutions and political agendas, especially nonreciprocal and extractive ones.

The global history of underwater mediation and augmentation at large cannot be decoupled from the history of European colonialism. Ocean historians including Helen Rozwadowski, Jon Crylen, and Nicole Starosielski have all noted, for instance, that films by Western ocean media-makers like Jacques Cousteau embedded colonial ambitions, positing "a stark distinction between nonwhite bodies, male and female, that adapted naturally to the undersea and the white male bodies of the cadets who conquered the ocean through technology."[6] In popular and scientific mediations, underwater technology has played into racializing processes and desires to "civilize" new frontiers. Given the damaging and often shortsighted motivations behind deep-ocean mediation, what is needed is a revised, decolonial perspective on how underwater mediation can contribute to broadening our prospects.

Decolonial and multispecies perspectives, which are what Macarena Gómez-Barris refers to as "submerged perspectives," help us think about mediation itself as a process that involves both the human and the more-than-human.[7] This has the effect of upending the naturalization of values such as continuity, coverage, and accuracy when it comes to technological enhancement. In chapter 5, I mentioned Shelley Denny's critique that the use of the term "integration" in reference to technical ocean-observing systems is often operationally a synonym for "assimilation."[8] We can extend this oppositional perspective further if we embrace and recognize how sensory augmentation and sensory deprivation participate in creating a multiplicity of mediated sensoriums. As long as dominant discourses of underwater technology maintain a perspective of human augmentation, we are not only excluding centuries of "low-tech" attentiveness to oceans but also continuing to perpetuate notions of being-in-the-world that center a very narrow band of the human sensorium—one that is myopic and ill-equipped to address environmental justice aims of protecting and living with nonhuman others.

Finally, "decolonizing" our mediations of the ocean requires not just representing animal life in our databases and legal processes or appealing to our sense of wonder and empathy, but also an ontological reorientation that centers experiences and knowledges outside of Western colonial frameworks of humanism and posthumanism altogether. This leads me to my second proposition for ethical mediation.

Build Offshore Humanisms

In closing this project, I find myself returning often to Paul Gilroy's proposition of an "offshore humanism," which Gilroy envisions as a notion of common belonging that articulates a shared vulnerability in salvage, in turbulence, and in planetary awareness. Writing about shipwrecked migrants, he explains, "Their salty saturation communicates something of the way that being human is transformed when the solidity of territory is left behind. We are afforded a glimpse of vulnerable, offshore humanity that might, in term, yield an *offshore* humanism."[9] Offshore humanisms, unlike terrestrial humanisms, are marked by operations beyond the bounds of sovereign power and bourgeois democracy. In addition to Gilroy's stories of courageous rescues of refugees, I increasingly see the interventions of this book and its emphasis on a shared sense of dependency and responsibility for ocean environments as contributing to a notion of offshore humanism that is bounded not by species lines or by borders, but by water itself.

My attraction to offshore humanism also has much to do with its centering of human transformations in the context of multispecies worlding. In response to the anthropocentrism inherent in mining, drilling, and other forms of resource extraction, environmental humanists have often turned to posthuman, nonhuman, multispecies, and inhuman agencies in their aspirations to construct an ethics of ecological responsibility. Scholarship from thinkers like James Bridle exemplify an underlying desire for a multispecies ethics in this disappearance of the human: "Ultimately, it's not about granting animals personhood, but about acknowledging and valuing their animalhood—and their planthood, their subjecthood, their beinghood."[10] Indeed, my own values in writing this book are largely aligned with this project. Multispecies belonging and beinghood are both highlighted in this book, and the presence of other species is core to my critique of extractive mediation.

However, I have reservations about turning to posthumanism as a means of describing this more-than-human awareness. Indigenous scholars have repeatedly argued that the philosophical push toward posthumanism is couched within settler philosophies, papering over or outright ignoring the thousands of years of indigenous cosmology and thought around human, animal, and environmental agencies that predate Eurocentric and postcolonial scholarship on the topic. Malreddy Pavan Kumar argues that this indigenous humanism is not merely reactionary against settler humanisms (as postcolonial humanisms often are), nor does it evacuate the specific landscape of harm and responsibility in the way that universalisms tend to. Citing the work of Canadian political philosopher Will Kymlicka as well as the 2007 United Nations Declaration on the

Rights of Indigenous Peoples, Kumar explains, indigenous peoples "do not seek a redemptive or restorative humanity. . . . Rather they seek a self-authorising humanity by means of: 1) the collectivity that is inherent to their cultures and 2) resistance to assimilationist forces that continue to threaten their traditional ways of living."[11] This collective humanism is rooted in an insistence that indigenous societies are irreducible to Western concepts of natural rights, as there is not a presumption of humanity's isolation from a mythical, Edenic nature in the first place.

This position is embedded in indigenous language around relations—not in the sense of a Spinozan "relational turn" that foregrounds language and relational subjectivity, but rather in the sense that we belong in both particular and multiple ways to all matter, and that animate matters have a familial relation of responsibility and kinship to each other. Sebastian De Line proposes the phrase "All my/our relations" to describe the decolonization of posthumanism: "it focuses on relationality while operating from within process and it makes this process personal, intimate and shared."[12] Framed through an indigenous cosmology, offshore humanism might be thought of as a practice—a methodology of being-in-relation to others without dispensing with our human bodies and human limitations.

Other ocean humanists have already taken up these threads, looking to the animal world for inspiration. Reflecting on Alexis Pauline Gumbs's cetacean tutelage and musings on breath, Astrida Neimanis asks, "How will those of us who have precipitated breathlessness—through domination, complicity, or carelessness—undergo our own transformations, to undo the versions of the 'human' that demanded these violences and extracted those breaths?"[13] From my perspective, Neimanis and Gilroy are seeking something in common—a way of saturating ourselves with a sense of our shared vulnerability with and responsibility to others, both human and nonhuman. In a similar vein, my turn toward whales, shrimp, and other undersea creatures in this book is not simply a reaction to the ills of anthropocentrism, but rather a grappling with Neimanis's question of what humanism can look like when it includes an environmental subjectivity.

The imaginaries I explored in chapter 4, which position the ocean as a breathing entity with a pulse, exist in some ways on a level of shallow analogy. Yet they also gesture toward shared affects and ruptures in material and temporal experience: the toxic entanglements of trash and treasure, the anticipatory mournings, and what Sarah Dimick calls the "climate arrythmias" of global warming—all of which are exacerbated by a voracious extractive capitalism that produces these erratic conditions.[14] Recognizing these common experiences of turbulence and, most importantly, their convergent institutional and ideological drivers across species lines is the first step in constructing an offshore humanism, as well as in revealing the mediated ecologies through which justice and injustice are sustained.

Finally, to speak of the something like the posthuman or nonhuman as diametrically opposed to anthropocentrism is to circumvent the deeply human origins of technological existence. Discourses of the posthuman tend to center the agencies of objects and of technology, but by doing so they run the risk of underplaying the human and political origins of the cybernetic social organization that now manifests in the "smart" ocean or Oceans 2.0. In fact, historians of computation such as Bernard Dionysius Geoghegan have explained that the "posthumanism" of cybernetics and its accompanying data-driven cultural adjustments were arranged not merely around technology but around the very idea of human enclosure and extraction—the colony, the asylum, and the camp.[15] Similarly, humanity was always already aquatic, long before we invented underwater drones, petroleum seismology, and underwater telepresence. Ocean mediation has never been a "posthuman" condition.

Attend to the Contact Zones of Mediation

Beyond centering offshore humanisms that foreground responsibility, ethical mediations can be animated by empathy and the ability to put ourselves in another's shoes (or perhaps another's suckers or antennae). An ideal sensing experience for a whale is not necessarily what is ideal for a human, and the same goes for microbes, crustaceans, and other critters of the sea. Who or what else is capable of bearing witness to the ocean and the struggles of the life within it?

In chapter 5, I introduced the "contact zone" to describe the unequal spaces of mediation through which humans and marine life encounter one another. Ocean observation benefits from an awareness of an oppositional, nonhuman gaze to our looks.[16] In the context of our oceans, that gaze can come from both human and nonhuman subjects. What does the glass octopus make of a fiber-optic undersea cable? What does a sea urchin sense in a sensor? How does a bamboo coral weather the storms of an underwater avalanche? How might a deep-sea shrimp react to a visiting AUV? Marine science and oceanographic tools can help us answer these questions—not just for the "benefit of mankind," but for our benthic brethren too. Prioritizing the needs and desires of ocean creatures is an important step toward thriving together.

It is not just whales and crustaceans that bear witness to the ocean's transformation. The elemental world registers these interactions as well. Contact zones like the AUV, the undersea cable, the ship hull, or the biofouled platform are interfaces—not just the kind composed of sensors or screens or other underwater media devices, but an always-changing boundary sedimented with ocean life, human labor, scientific knowledge, political agendas, and more. The materiality of the contact zone broadens who and what we validate as "observers" in the first place. Consider, for instance, how the biofouled cable acts as a witness to both the ocean's health and the transformations of the cable itself as it binds with its environment. Benthic organisms with bodies filled with toxins are also "material witnesses" to extraction and its excesses, providing scientists with evidence of harm.[17] Tracer dyes reproduce this evidentiary event in a less toxic but more perceptible form. And while Gilroy speaks of bearing witness to the shipwreck, the shipwreck itself is its own material witness to colonialism, oceanic turbulence, and the ever-present ecological flows that make up the very media of life. The contact zone is an archive that exhibits changes to the ocean itself, a lively and multispecies worlding.

My conflations here between the contact zone, the archive, and the interface emphasize both the spatiality of mediation and the multiple temporalities through which we observe the ocean and through which the ocean observes us. Susan Schuppli contends that the "material witness" alternates from "divulging 'evidence of the event' and exposing the 'event of evidence.'"[18] This is to say that matter does not bear witness equally and that witnessing is contingent, "soliciting questions about what can be known in relationship to that which is seen or sensed, about how or what is able to bestow meaning onto things, and about whose stories will be heeded or dismissed."[19] I share this sense of contingency in my analysis of oceanographic texts like environmental impact statements, where information is unevenly recruited as evidence of acceptable or unacceptable harm. One could argue that justifications for continuing extractive developments at the seafloor on such a massive scale signal how, for many developers, the seafloor "does not matter" at all.

There is an overarching suggestion, sometimes articulated explicitly and sometimes implicitly by ocean scientists and policymakers, that destroying the seafloor, which is seldom seen by ordinary people, is preferable to destroying terrestrial landscapes, which are seen by humans on a day-to-day basis. Such ocularcentric logic blatantly ignores our ecological connection to the oceans and its ongoing necessity to human survival. It is at our peril that we choose to defile the deep seabed, a key producer and distributor of food, nutrients, oxygen, and other materials for the human and nonhuman worlds around the planet. When we are told to choose the lesser of two evils, we are given a misleading choice of this or that, ignoring all other options that imagine sustainability outside of an ideology of technological growth and profit. Despite these challenges, I believe that a renewed perspective from the contact zone and from the material witnesses of anthropogenic disturbance—whether that is extraction or extractive mediation—acts as a defense against oceanic erasures.

Make Space for the Unknown

Perhaps one of the most difficult—yet also most important—conclusions that can be made from my research here is that, despite the widespread perception that more knowledge leads to less abuse, mediation also produces vulnerability that can at times be deleterious. Of all the ocean stakeholders and scientists I talked to over the past five years for this project, I only encountered one person who even broached the question of what it would mean to resist the urge to collect data. This was an indigenous knowledge holder who, during a panel session at Ocean-Obs'19, raised the point that in his experience, when access is given to ocean scientists to observe a local environment, that access is quickly exploited and often results in the eventual destruction of that ecosystem. The comment was about extractive mediation, and the lack of coherent response to his question at that time has been at the back of my mind ever since I undertook this project.

Too often, our dearth of knowledge about the deep sea has been extrapolated onto an imaginary of the deep as void, as desert, and as dead. Evidence of the deep sea's abundance from dredging and other historical expeditions was often framed through this expectation of void, and as Stacy Alaimo describes, that disjunct produced monstrous imaginaries of deep-sea dwellers, divorced from the unmoored and submerged "violet-black ecologies" of the benthos itself.[20] Taking into account these discursive constructions of voids and knowledge gaps, it is clear that human relationships to deepwater ecosystems are produced both by what we observe and by that which we do not yet have the capacity to understand. And while the question of fidelity in representation is privileged in scientific mediations of the ocean, mediation itself is never strictly a matter of fidelity—it is also a creative process that enables us to explore our positionality in relation to our environments.

Artists have often used aesthetic engagements to reflect on the mingled affects of fear and curiosity that our ocean encounters engender. I think of the insights shared, for instance, by Sofia Crespo, a media artist at the Entangled Others Studio: "To tell a story is to be confronted with what we know, what we believe, and the unknown."[21] Every mediation that we create of our oceans is a confrontation of sorts—an attempt to arrange and make sense. When we visualize what we know, we are simultaneously visualizing the limits to our knowledge, and the thresholds between knowing and not knowing. But we need not fear these unknown territories, framing them as gaps or as frontiers to assuage our anxieties. Caring for our oceans is contingent on our "staying with the trouble" and acknowledging the unfathomable parts of ocean existence.[22]

In chapter 5, I mentioned a VR simulation of a deep-sea future created by the U.S. Department of Energy, which depicts a busy seabed full of underwater turbines, AUVs, and aquaculture facilities. As a counterpart to this vision of our ocean future, I want to turn to a few examples of deep-sea art that embed not the mandate for clear images and data but rather a charge to depict real feelings, relationships, and processes that structure human relation to the deep ocean, leaving space for imagination and for the unknown. One such person is Portuguese animator Alice Aires, who imagines something quite unique in her speculative bioluminescent ecology. Aires draws from scientific aesthetics, deep-sea biology, and her own concerns about the ocean pollution in her animated short *Plastified: An Ode* (see Figure 22). Her website describes *Plastified*'s

Figure 22. A still from *Plastified: An Ode*, a short film by Alice Aires, depicting a bioluminescent marine invertebrate resembling a sea anemone. Image by Alice Aires.

Figure 23. In Other Waters relies on a minimalist navigation interface to signify alien, underwater life, rather than photorealistic depictions of creatures. Copyright Gareth Damian Martin, Jump Over the Age.

virtual deep-sea dive into the twilight zone as a "partially speculative ecology" in which "a food chain of bioluminescent creatures is digitally memorialized by the cause of their extinction—plastic." The result is part adventure, part science, and part eulogy. Though we are given images of glowing microbes and tentacles and electrified corals not known to humans, there is a realism to these flows of image and sound, which were created in consultation with scientist Inês Ferreira Guedes.[23] Produced with tidbits of real knowledge, it is nevertheless an ode to the not-yet-known and perhaps never-will-be-known abundance of the deep. Other deep-sea animations, including games like Jump Over the Age's *In Other Waters* (2020), eschew the expectation for a transparent, submerged perspective on the deep sea altogether (see Figure 23). *In Other Waters* is notable for the way it opts for a high-tech yet minimalist navigation interface, which reminds its players that we can never separate our understanding of the deep sea from our media tools. Despite its antivisual premise, this text-heavy science fictional adventure is full of realistic and immersive details, as players must collect samples, interact with AI machines, and ultimately confront their own ethical and moral responsibility toward an alien ocean ecosystem as scientists employed by a fictional mining corporation. More than showing us a look of discovery, media about the deep sea can provide occasions for us to reflect on how we come to knowledge and where the limits of that knowledge are. We may never know how the shrimp perceives the drone, or how the whale feels its tag. But if we aim to foster relations of care with aquatic life through mediation, perhaps the lesson we can take with us is that caring is not merely a question of curing, nor is it reducible to mere physical proximity. If these things were true, then there would be no reason for anyone living in the middle of a continent to care about what happens on the coasts. There would be no impetus for a bad swimmer from Colorado to grow up and write a book about the deep sea.

ACKNOWLEDGMENTS

What a journey it has been to write this book. I remember when the seeds of this project were germinating in the TA offices and classrooms of UC Santa Barbara, where as a graduate student I first felt the pull of the Pacific Ocean. I started playing with ideas around oceanic media, sonic environments, and energy futures on the California coast, and each subsequent encounter, from the blustery harbor at Woods Hole to the dark waters of Pearl Harbor, percolated into the layers of this book.

I want to start by expressing my sincere gratitude to my UCSB mentors. I owe a special thanks to my PhD adviser and mentor, Alenda Chang, for her steadfast encouragement and support throughout my academic journey, and to Melody Jue, whose brilliant method of submerging media concepts underwater forever changed the way I thought about the field of media studies. I offer thanks also to Janet Walker, Lisa Parks, Bhaskar Sarkar, Bishnupriya Ghosh, Cristina Venegas, Chuck Wolfe, Tyler Morgenstern, and David Novak, who so generously lent their time to read my early work and bounce off ideas with me. This research was also made possible by the generous financial support of the Graduate Humanities Research Fellowship at UCSB as well as funding from the ASU Humanities Institute.

Next, I want to acknowledge my colleagues at Arizona State University and beyond who helped me take this book into its final phase. Thank you to editor Leah Pennywark at the University of Minnesota Press for taking an interest in this book. A big thank you also goes out to Natalie Lozinski-Veach, Dan Gilfillan, and Jacob Greene for offering feedback on my chapters, as well as to John Balch for talking alchemy with me and giving me the best recommendations. I am so grateful to the ASU

Faculty Women of Color Caucus for providing a daily space for writing and encouragement. To Katie Morrissey, Amaru Tejeda, Rachel Fabian, and Bianka Ballina, moving to Arizona and writing this book during the pandemic was not easy, and our Zoom coworking sessions kept me sane. I am also thankful for Joni Adamson with the ASU Environmental Humanities Initiative and Ron Broglio at the ASU Humanities Institute for creating rich spaces and opportunities at ASU for my work to flourish.

I feel very lucky to be part of a community of environmental humanists, ocean researchers, and critical media scholars who inspire me every day with their bold ideas and the sense of care that they put into their work. My gratitude goes out to all those who have kindly invited me to workshops, panels, and publications and other occasions to be curious and get excited about new ideas, reminding me just what it is I love about academic research: Jen Telesca, Justine Bakker, Bernard Geoghegan, Anne Pasek, Thomas Pringle, Yuriko Furuhata, Derek Woods, Jordan Kinder, Josef Nguyen, Stefan Helmreich, Nicole Starosielski, Rafico Ruiz, John Shiga, Jess Lehman, Chris Walker, Tamara Fernando, Christina Vagt, Wolf Kittler, Sara Rich, Elizabeth DeLoughrey, Stacy Alaimo, Jeff Scheible, Karen Redrobe, Weixian Pan, Patrick Brodie, Carlos Alonso Nugent, Andrea Ballestero, Helen Rozwadowski, Celina Osuna, Mako Fitts Ward, Sage Gerson, Pujita Guha, Stephen Borunda, and many more. I would like to add a special thanks to the late Sol Neely as well, who helped transform my understanding of Indigenous environmental philosophy.

It would not have been possible to write this book without the occasion to learn from ocean scientists and engineers, and I am so grateful to those who have graciously taken the time to talk to me about their research: Elizabeth Burgess, Daniel Palacios, Elizabeth Becker, Veevee Cai, Dillon McEwan, Alexandra Hangsterfer, Jim Broda, Val Stanley, Joe Stoner, Deborah Kelley, Eckert Meiburg, Caroline Jablonicky, Cindy Dover, Nick Pisias, Maziet Cheseby, Paul Walczak, and the archivists at Woods Hole Oceanographic Institution and MIT Special Collections. I have so much respect for all that you do to protect and understand ocean environments. Your kindness, sincerity, and willingness to engage across disciplinary fields truly give me hope for the future.

Finally, I thank my friends and family for providing safe spaces to process all the craziness of the past several years. I do not know that I would be where I am without the friendship, solidarity, and sense of community that you helped create. I thank my friends Lauren Mahoney, Gordon Kirby, Erick Rodriguez, and Alex Champlin for the emotional support; and my NCFDD small group, Lisa McManus, Joyce He, and Zibei Chen, for the additional accountability. Thank you to my brother, Leo, for always being there to give me perspective, and to my parents, Min and Yuming, for the support. And finally, to my partner, Johnny, words cannot express the depth of my gratitude for your positivity and companionship, in the hard times and the good. Thank you.

NOTES

Introduction

1. Julie Sze, *Fantasy Islands: Chinese Dreams and Ecological Fears in an Age of Climate Crisis* (Oakland: University of California Press, 2015).

2. Claiborne Pell, "The Scramble Is On for Ocean Riches," *The World*, November 12, 1967, 21, MC-16, box 29, folder 3, Columbus O'Donnell Iselin Papers, Woods Hole Data Library and Archives, Woods Hole, Massachusetts.

3. Ann Elias, *Coral Empire: Underwater Oceans, Colonial Tropics, Visual Modernity* (Durham: Duke University Press, 2019), 96.

4. Elias, 21–22.

5. William Newman, *Promethean Ambitions: Alchemy and the Quest to Perfect Nature* (Chicago: University of Chicago Press, 2004), 25; Donna Haraway, *The Companion Species Manifesto: Dogs, People, and Significant Otherness*, vol. 1 (Chicago: Prickly Paradigm Press, 2003).

6. Deborah James, "The Alchemy of a Corpus of Underwater Images: Locating Carysfort to Reconcile our Human Relationship with a Coral Reef," *Nature + Culture* 15, no. 3 (December 2020): 244.

7. Pamela Smith, *The Business of Alchemy: Science and Culture in the Holy Roman Empire* (Princeton: Princeton University Press, 1994), 8–9.

8. Katherine Sammler, "The Deep Pacific: Island Governance and Seabed Mineral Development," in *Island Geographies: Essays and Conversations*, ed. Elaine Stratford (London: Routledge Studies in Human Geography, 2017), 13–14.

9. "High Seas," *Encyclopaedia Britannica*, https://www.britannica.com/topic/high-seas#ref215901.

10. See Saskia Sassen, *Expulsions: Brutality and Complexity in the Global Economy* (Cambridge, Mass.: Harvard University Press, 2014).

11. Jeffrey A. Karson et al., *Discovering the Deep: A Photographic Atlas of the Seafloor and Ocean Crust* (Cambridge, UK: Cambridge University Press, 2015), ixx.

12. See Adrienne Buller, *The Value of a Whale: On the Illusions of Green Capitalism* (Manchester: Manchester University Press, 2022).

13. Joshua Ramey, *Politics of Divination: Neoliberal Endgame and the Religion of Contingency* (Lanham, Md.: Rowman and Littlefield, 2016); John Zysman and

Mark Huberty, *Can Green Sustain Growth? From the Religion to the Reality of Sustainable Prosperity* (Stanford: Stanford University Press, 2013).

14. Elizabeth DeLoughrey, "Mining the Seas: Speculative Fictions and Futures," in *Laws of the Sea: Interdisciplinary Currents*, ed. Irus Braverman (London: Routledge, 2022), 146.

15. Raymond Williams, "Mediation," in *Keywords: A Vocabulary of Culture and Society* (New York: Oxford University Press, 1976), 206–7.

16. Jonathan Beller, "The Cinematic Mode of Production," *Culture, Theory & Critique* 44, no. 1 (2003): 95.

17. Stuart Hall, "Representation and the Media" (lecture, the Open University, Media Education Foundation, 1997), 14.

18. Steve Mentz, "Toward a Blue Cultural Studies: The Sea, Maritime Culture, and Early Modern English Literature," *Literature Compass* 6, no. 5 (2009): 997–1013; Stacy Alaimo, "Introduction: Science Studies and the Blue Humanities," *Configurations* 27, no. 4 (2019): 429–32.

19. Kim De Wolff, Rina C. Faletti, and Ignacio López-Calvo, *Hydrohumanities: Water Discourse and Environmental Futures* (Oakland: University of California Press, 2021).

20. DeLoughrey, "Mining the Seas," 146.

21. Lisa Parks, "'Stuff You Can Kick': Conceptualizing Media Infrastructures," in *Humanities and the Digital*, ed. David Theo Goldberg and Patrik Svensson (Cambridge: MIT Press, 2015), 355–74. Also see Shannon Mattern, "Mission Control: A History of the Urban Dashboard," *Places Journal*, March 2015, https://doi.org/10.22269/150309.

22. Nicole Starosielski, *The Undersea Network* (Durham: Duke University Press, 2015).

23. Lisa Gitelman, "Holding Electronic Networks by the Wrong End," *Amodern 2: Network Archaeology*, October 2013, https://amodern.net/article/holding-electronic-networks-by-the-wrong-end/.

24. Steven J. Jackson, "Rethinking Repair," in *Media Technologies: Essays on Communication, Materiality, and Society*, ed. Tarleton Gillespie, Pablo J. Boczkowski, and Kirsten A. Foot (Cambridge: MIT Press, 2014), 221–39.

25. See Jennifer Gabrys, *Digital Rubbish: A Natural History of Electronics* (Ann Arbor: University of Michigan Press, 2011).

26. John Durham Peters, *The Marvelous Clouds: Towards an Elemental Theory of Media* (Chicago: University of Chicago Press, 2015), 3.

27. Melody Jue, *Wild Blue Media* (Durham: Duke University Press, 2020).

28. Bernhard Siegert, *Cultural Techniques: Grids, Filters, Doors, and Other Articulations of the Real*, trans. Geoffrey Winthrop-Young (New York: Fordham University Press, 2015).

29. Sarah Kember and Joanna Zylinska, *Life after New Media: Mediation as a Vital Process* (Cambridge: MIT Press, 2012), xv.

30. Richard Grusin, "Radical Mediation," *Critical Inquiry* 42, no. 1 (Autumn 2015): 132.

31. Donna Haraway, "Situated Knowledges: The Science Question in Feminism and the Privilege of Partial Perspective," *Feminist Studies* 14, no. 3 (1988): 579.

32. S. Eben Kirksey and Stefan Helmreich, "The Emergence of Multispecies Ethnography," *Cultural Anthropology* 25, no. 4 (2010): 545–76, https://doi.org/10.1111/j.1548-1360.2010.01069.x.

33. Ursula Heise, *Imagining Extinction: The Cultural Meanings of Endangered Species* (Chicago: University of Chicago Press, 2016), 167.

34. Anna Tsing, *Friction: An Ethnography of Global Connection* (Princeton: Princeton University Press, 2004), 176.

35. Naomi Klein, *This Changes Everything: Capitalism vs. the Climate* (New York: Simon and Schuster, 2014), 169.

36. Jordan Kinder, "Gaming Extractivism: Indigenous Resurgence, Unjust Infrastructures, and the Politics of Play in Elizabeth LaPensée's *Thunderbird Strike*," *Canadian Journal of Communication* 46, no. 2 (April 2021): 247–69.

37. Macarena Gómez-Barris, *The Extractive Zone: Social Ecologies and Decolonial Perspectives (Dissident Acts)* (Durham: Duke University Press, 2017), xvi.

38. Lowell Duckert, "Earth's Prospects," in *Elemental Ecocriticism*, ed. Jeffrey Jerome Cohen and Lowell Duckert (Minneapolis: University of Minnesota Press, 2015), location 5119 of 7797, Kindle.

39. Brian Jacobson, "Prospecting: Cinema and the Exploration of Extraction," in *Cinema of Exploration: Essays on an Adventurous Film Practice*, ed. James Leo Cahill and Luca Caminati (New York: Routledge, 2020), 283.

40. Frantz Fanon, *Black Skin, White Masks*, trans. Charles Lam Markmann (London: Pluto Press, 1986), 15.

41. A. Pardo—Interventions, Papers, 1973, 03–041, box 5, United Nations Conference on the Law of the Sea Collection and Related Materials, 1938–1982, University of Washington Libraries, Special Collections, Seattle, Washington.

42. Lisa Yin Han, "Transparency at Depth: Dark Mediation of the Deep Seabed," in *Deep Mediations: Thinking Space in Cinema and Digital Cultures*, ed. Karen Redrobe and Jeff Scheible (Minneapolis: University of Minnesota Press, 2020), 242–43.

43. James Hamilton-Paterson, "Wrecks and Death," in *Seven-Tenths: The Sea and Its Thresholds* (London: Faber and Faber, 2007), 179.

44. See Melody Jue and Rafico Ruiz, *Saturation: An Elemental Politics* (Durham: Duke University Press, 2021).

45. Elizabeth M. DeLoughrey, *Allegories of the Anthropocene* (Durham: Duke University Press, 2019), 7.

46. Michel Callon, "Some Elements of a Sociology of Translation: Domestication of the Scallops and the Fishermen of St. Brieuc Bay," in *Technoscience: The Politics of Interventions*, ed. Kristin Asdal, Brita Brenna, and Ingunn Moser (N.p.: Oslo Academic Press, 2007), 57–78.

47. Susan Leigh Star and James R. Griesemer, "Institutional Ecology, 'Translations,' and Boundary Objects: Amateurs and Professionals in Berkeley's

Museum of Vertebrate Zoology, 1907–39." *Social Studies of Science* 19, no. 3 (1989): 389–90.

48. Michel Foucault, *The Archaeology of Knowledge and the Discourse on Language, trans. Alan Sheridan* (New York: Pantheon Books, 1972), 129–30.

49. Michel Foucault, "Genealogy, Nietzsche, History," in *The Foucault Reader*, ed. Paul Rabinow (New York: Pantheon Books, 1984), 77.

50. Foucault, *Archaeology of Knowledge.*

51. Jason Groves, "An Anthropocene Observatory," *Open Humanities Press*, March 4, 2016, http://openhumanitiespress.org/feedback/newecologies/anthropocene_observatory/.

1. The Blue Archive and the Blue Frontier

1. Steve Mentz, *Shipwreck Modernity: Ecologies of Globalization, 1550–1790* (Minneapolis: University of Minnesota Press, 2015), location 275 of 5584, Kindle.

2. Jay Bennett, "Less than 1 Percent of the World's Shipwrecks Have Been Explored," *Popular Mechanics*, January 18, 2016, http://www.popularmechanics.com/science/a19000/less-than-one-percent-worlds-shipwrecks-explored/.

3. Anslem L. Strauss, *Negotiations: Varieties, Contexts, Processes, and Social Order* (San Francisco: Jossey-Bass, 1978).

4. James Hamilton-Paterson, "Wrecks and Death," in *Seven-Tenths: The Sea and Its Thresholds* (London: Faber and Faber, 2007), 154.

5. Hamilton-Paterson, 161.

6. Sara Rich, *Shipwreck Hauntography: Underwater Ruins and the Uncanny* (Amsterdam: Amsterdam University Press, 2021).

7. Peter Campbell, "Could Shipwreck Lead the World to War?," *New York Times*, December 18, 2015, https://www.nytimes.com/2015/12/19/opinion/could-shipwrecks-lead-the-world-to-war.html.

8. Roland Oliphant, "Vladimir Putin Plunges into Black Sea in Research Submarine," *The Telegraph*, August 18, 2015, https://www.telegraph.co.uk/news/worldnews/europe/russia/11810703/Vladimir-Putin-plunges-into-Black-Sea-in-research-submarine.html.

9. Nadia Abu El-Haj, *Facts on the Ground: Archaeological Practice and Territorial Self-Fashioning in Israeli Society* (Chicago: University of Chicago Press, 2001), 9.

10. Naor Ben-Yehoyada, "Heritage Washed Ashore: Underwater Archaeology and Regionalist Imaginaries in the Central Mediterranean," in *Critically Mediterranean: Temporalities, Aesthetics, and Deployments of a Sea in Crisis*, ed. yasser elhariry and Edwige Tamalet Talbayev (New York: Palgrave Macmillan, 2018), 229.

11. Joan Scott, "Fantasy Echo: History and the Construction of Identity," *Critical Inquiry* 27 (Winter 2001): 290.

12. Scott, 292.

13. While originally referring to the remembrance offered by museums, the term "musealization" as used by Huyssen describes an "expansive historicism of our contemporary culture, a cultural present gripped with an unprecedented

obsession with the past." Andreas Huyssen, "Present Pasts: Media, Politics, Amnesia," *Public Culture* 12, no. 1 (Winter 2000): 32.

14. Greg Siegel, *Forensic Media: Reconstructing Accidents in Accelerated Modernity* (Durham: Duke University Press, 2014), 28.

15. "Titanic," *History*, https://www.history.com/topics/early-20th-century-us/titanic.

16. Submarine Signal Company, "The Development of the Fathometer and Echo Depth Finding," *Soundings*, April 11, 1932.

17. UNESCO, "The Guangdong Maritime Silk Road Museum (Nanhai No. 1 Museum), Yangjiang, Guangdong Province, China," Underwater Cultural Heritage, http://www.unesco.org/new/en/culture/themes/underwater-cultural-heritage/about-the-heritage/underwater-museums/the-guangdong-maritime-silk-road-museum-nanhai-no-1-museum/.

18. UNESCO.

19. Yongjie Xu, "The Test Excavation of the Nanhai No. 1 Shipwreck in 2011: A Detail Leading to the Whole," *The Silk Road* 13 (2015): 84.

20. Xu, 87.

21. Liu Jin, "Sisha 2015 Underwater Archaeology Has Officially Set Sail," State Administration of Cultural Heritage, April 13, 2015, http://www.sach.gov.cn/art/2015/4/13/art_722_118859.html; my translation.

22. This practice of using marine archaeology as a political maneuver also resounds with the nation's hotly contested island building initiatives, another way in which narrative and physical control over ocean space is associated with geopolitical power. Julie Sze, *Fantasy Islands: Chinese Dreams and Ecological Fears in an Age of Climate Crisis* (Berkeley: University of California Press, 2015).

23. Jean-Marc F. Blanchard and Colin Flint, "The Geopolitics of China's Maritime Silk Road Initiative," *Geopolitics* 22, no. 2 (2017): 223–45, https://doi.org/10.1080/14650045.2017.1291503.

24. "Maritime Silk Road Museum of Guangdong," Trip Advisor Reviews, https://www.tripadvisor.com/Attraction_Review-g659643-d5827287-Reviews-Maritime_Silk_Road_Museum_of_Guangdong-Yangjiang_Guangdong.html; Maritime Silk Road Museum of Guangdong, https://www.msrmuseum.com/Home/Enindex.

25. Mark O'Neill, "The Return of Nanhai No. 1," *Macao Magazine*, July 3, 2018, https://www.macaomagazine.net/the-return-of-nanhai-no-1.

26. See Wendy Hui Kyong Chun, "The Enduring Ephemeral, or the Future Is a Memory," *Critical Inquiry* 35, no. 1 (2008): 148–71, https://www.journals.uchicago.edu/doi/10.1086/595632; Wolfgang Ernst, "Dis/continuities: Does the Archive Become Metaphorical in Multi-media Space?," in *New Media Old Media: A History and Theory Reader*, ed. Wendy Hui Kyong Chun, Anna Watkins Fisher, and Thomas Keenan (New York: Routledge, 2006), 105–23; Ann Stoler, *Along the Archival Grain: Epistemic Anxieties and Colonial Common Sense* (Princeton: Princeton University Press, 2010).

27. Alenda Chang, "Environmental Remediation," *Electronic Book Review*, June 7, 2015, 5.

28. Édouard Glissant, *Poetics of Relation*, trans. Betsy Wing (Ann Arbor: University of Michigan Press, 1997).

29. Bennett Hall, "At the Earth's Core," October 5, 2015, *Corvallis Gazette-Times*. https://www.gazettetimes.com/news/local/at-the-earth-s-core/article_ff 51b324-ef97-501c-8c14-277df9a03909.html.

30. Joseph Stoner, interview with author, Oregon State University, August 16, 2018.

31. Elizabeth A. Povinelli, Matthew Coleman, and Katheryn Yusoff, "An Interview with Elizabeth Povinelli: Geontopower, Biopolitics, and the Anthropocene," *Theory, Culture & Society* 34, nos. 2–3 (2017): 175.

32. Elizabeth Grosz, Kathryn Yusoff, and Nigel Clark, "An Interview with Elizabeth Grosz: Geopower, Inhumanism, and the Biopolitical," *Theory, Culture & Society* 34, nos. 2–3 (2017): 131.

33. Maritime Silk Road Museum of Guangdong.

34. See Zoe Todd, "Fossil Fuels and Fossil Kin: An Environmental Kin Study of Weaponised Fossil Kin and Alberta's So-Called 'Energy Resources Heritage,'" *Antipode* (2023): 1–25, https://doi.org/10.1111/anti.12897.

35. El-Haj, *Facts on the Ground*, 13.

36. "About," Institute of Nautical Archaeology, 2018, https://nauticalarch.org/.

37. Ben-Yehoyada explains that "the present constellation combines *similarities* to the sea's previous lives and the *continuities* that people draw from ancient maritime pasts they access and the transnational present they inhabit." Ben-Yehoyada, "Heritage Washed Ashore," 220.

38. Jacques Derrida, *Archive Fever: A Freudian Impression* (Chicago: University of Chicago Press, 1996), 17; Ernst, "Dis/continuities."

39. Derrida, 17.

40. Lucy Suchman, *Human–Machine Reconfigurations: Plans and Situated Actions* (New York: Cambridge University Press, 2007), 263. Suchman's ideas on expertise resound with arguments by Bruno Latour, who has similarly argued for reflexivity about the cultural assumptions and social negotiations within the laboratory. See Bruno Latour and Steve Woolgar. *Laboratory Life: The Construction of Scientific Facts* (Princeton: Princeton University Press, 2013); Susan Leigh Star and James R. Griesemer, "Institutional Ecology, Translations and Boundary Objects: Amateurs and Professionals in Berkeley's Museum of Vertebrate Zoology, 1907–39." *Social Studies of Science* 19, no. 3 (1989): 387–420.

41. Lisa Parks, *Cultures in Orbit: Satellites and the Televisual* (Durham: Duke University Press, 2005): 110, 114.

42. Parks, 112.

43. Parks, 111.

44. Melody Jue, "Proteus and the Digital: Scalar Transformations of Seawater's Materiality in Ocean Animations," *animation* 9, no. 2 (2014): 246.

45. Matt Bardo, "The Wild World of Shipwrecks," *BBC Nature*, April 20, 2012, http://www.bbc.co.uk/nature/17706609.

46. David Lowenthal, "Material Preservation and Its Alternatives," *Perspecta* 25 (1989): 69, https://doi.org/10.2307/1567139.

47. In the field of film preservation, preservation might include the process of restoring content and protecting its integrity, or it might involve minimizing degradation. See Paolo Cherchi Usai, "The Ethics of Film Preservation," in *Silent Cinema: An Introduction* (London: BFI, 2000), 66.

48. For more on conservation techniques, see Donny L. Hamilton, *Methods of Conserving Archaeological Material from Underwater Sites* (Department of Anthropology, Texas A&M University, 1999), http://nautarch.tamu.edu/CRL/conservationmanual/ConservationManual.pdf.

49. "Clan Ranald Anchor Returns to Edithburgh after Restoration Work," *ABC*, September 22, 2014, http://www.abc.net.au/news/2014-09-22/clan-ranald-anchor-returns-home-after-lengthy-restoration/5759930.

50. Hiram Morgan, "A Race against Time to Save Spanish Armada Wrecks before They Are Lost Forever," *The Irish Times*, April 14, 2015, http://www.irishtimes.com/opinion/a-race-againsttime-to-save-spanish-armada-wrecks-before-they-are-lost-forever-1.2174364.

51. Elizabeth DeLoughrey. "Submarine Futures of the Anthropocene," in "Oceanic Routes Forum," special issue of *Comparative Literature Journal* 69, no. 1 (2017): 36, https://doi.org/10.1215/00104124-3794589.

52. Nautical Archaeology Society, "Policies and Statements," http://www.nauticalarchaeologysociety.org/content/policies-and-statements.

53. Michelle Barron, "Drowned in Law: An Examination of M. NourbeSe Philip's *Zong!* and the Regulation of Human Remains in International Waters," in *Underwater Worlds: Submerged Visions in Science and Culture*, ed. Will Abberley (Newcastle upon Tyne: Cambridge Scholars Publishing, 2018), 160.

54. Anastasia Strati, "Deep Seabed Cultural Property and the Common Heritage of Mankind," *International and Comparative Law Quarterly* 40, no. 4 (1991): 881; also see Anne M. Cottrell, "The Law of the Sea and International Marine Archaeology: Abandoning Admiralty Law to Protect Historic Shipwrecks," *Fordham International Law Journal* 17 no. 3 (1993): 667–725.

55. My emphasis. UN General Assembly, Article 149, *Convention on the Law of the Sea*, December 10, 1982; Lowell Bautista, "Ensuring the Preservation of Submerged Treasures for the Next Generation: The Protection of Underwater Cultural Heritage in International Law," *LOSI Conference Papers* (2012), https://www.law.berkeley.edu/files/Bautista-final.pdf.

56. Strati, "Deep Seabed Cultural Property," 879.

57. Jody Berland, *North of Empire: Essays on the Cultural Technologies of Space* (Durham: Duke University Press, 2009), 14.

58. Frederick Jackson Turner, *The Significance of the Frontier in American History* (London: Penguin UK, 2008), 3.

59. Sandro Mezzadra and Brett Nielson's work on *Border as Method* references this imbrication with the phrase, "primitive accumulation of modern cartography," which gestures toward the mutual production of capital and geographic

border zones. Sandro Mezzadra and Brett Nielson, *Border as Method, or, the Multiplication of Labor* (Durham: Duke University Press, 2013).

60. Antje Boetius and Matthias Haeckel, "Mind the Seafloor," *Science* 359, no. 6371 (January 5, 2018): 34, https://www.science.org/doi/10.1126/science.aap7301.

61. James Cameron, dir., *DeepSea Challenge* (National Geographic, 2014); Stefan Helmreich, *Alien Ocean: Anthropological Voyages in Microbial Seas* (Berkeley: University of California Press, 2009); Jacques Cousteau, *The Silent World: A Story of Undersea Discovery and Adventure* (Los Angeles: Columbia Pictures, 1953); Julian Smith, "Seafloor Miners Poised to Cut into an Invisible Frontier," *Scientific American*, August 11, 2016, https://www.scientificamerican.com/article/seafloor-miners-poised-to-cut-into-an-invisible-frontier/; Laura Hampton, "Deep-Sea Alliance Set to Probe Earth's Final Frontier," *New Scientist*, July 5, 2016, https://www.newscientist.com/article/2096187-deep-sea-alliance-set-to-probe-earths-final-frontier/; "Deep Sea Is the New Frontier," *Outside*, December 7, 2017, https://www.outsideonline.com/2266756/deep-sea-new-frontier.

62. UN General Assembly, Article 149, *Convention on the Law of the Sea*, December 10, 1982.

63. International Seabed Authority, "Exploitation," 2017, https://www.isa.org.jm/exploitation.

64. International Seabed Authority, Decision of the Assembly of the International Seabed Authority Relating to the Regulations on Prospecting and Exploration for Cobalt-Rich Ferromanganese Crusts in the Area, Eighteenth session, 138th meeting (Kingston, Jamaica, July 27, 2012), 6, https://www.isa.org.jm/mining_code/2012-isba-18-a-11/.

65. Elizabeth DeLoughrey, "Ordinary Futures: Interspecies Worldings in the Anthropocene," in *Global Ecologies and the Environmental Humanities: Postcolonial Approaches*, ed. Elizabeth DeLoughrey, Jill Didur, and Anthony Carrigan (New York: Routledge, 2015), 355.

66. DeLoughrey, 367.

67. "The Ocean Eventual Solution to Many Problems," *Encyclopedia of Marine Resources*, January 26, 1968, 9, Columbus O'Donnell Iselin Papers, 1904–1971, MC-16, box 4, Woods Hole Oceanographic Institution, Data Library and Archives (hereafter WHOI), Woods Hole, Massachusetts, http://archives.mblwhoilibrary.org:8081/repositories/2/resources/153.

68. Richard White, "Frederick Jackson Turner and Buffalo Bill," in *The Frontier in American Culture*, ed. James R. Grossman (Berkeley: University of California Press, 1994), 7–66.

69. Anna Tsing, "Natural Resources and Capitalist Frontiers," *Economic and Political Weekly* 38, no. 41 (2003): 5102.

70. Patricia Nelson Limerick, "The Adventures of the Frontier in the Twentieth Century," in *The Frontier in American Culture*, ed. James R. Grossman (Berkeley: University of California Press, 1994), 88.

71. "WHOI History during the War Years 1941–50," MC-16, box 31, folder 10, Iselin Papers, WHOI, http://archives.mblwhoilibrary.org:8081/repositories/2/resources/153.

72. Claire Calcagno, "Shipwreck Studies," The Edgerton Digital Collections Project, Massachusetts Institute of Technology, http://edgerton-digital -collections.org/stories/features/fathoming-the-oceans-8-shipwreck-studies.

73. Wendy Hui Kyong Chun, "The Enduring Ephemeral, or the Future Is a Memory," *Critical Inquiry* 35, no. 1 (2008): 50.

74. Paul Virilio, "The Primal Accident," in *The Politics of Everyday Fear*, ed. Brian Massumi (Minneapolis: University of Minnesota Press, 1993), 212.

75. Bruno Latour, "Visualization and Cognition," *Knowledge and Society* 6, no. 6 (1986): 1–40.

76. Imagery typically comes from ROVs hooked up to high-definition cameras that stream images via fiber-optic cable directly to control rooms. This is then sent via satellite to receiving stations, and then distributed over the web. Nautilus Live, "The Tools of Ocean Exploration," https://nautiluslive.org/tech.

77. Megan Chen, David Downing, and Linda Fergusson-Kolmes, "Rediscovering History: Submarine USS Bugara," Nautilus Live, August 25, 2017, https://nautiluslive.org/album/2017/08/28/rediscovering-history-submarine -uss-bugara.

78. Damien Hirst, *Treasures from the Wreck of the Unbelievable*, dir. Sam Hobkinson (Netflix, 2017); Ruth Wallen, "The Sea as Sculptress—From Analog to Digital," *UC Irvine: Digital Arts and Culture 2009*, https://escholarship .org/uc/item/3pm5b4jp; "Home," Museo Subacuático de Arte, 2021, https:// musamexico.org/.

79. Mentz, *Shipwreck Modernity*, locations 79–83 of 5584, Kindle.

80. Dominik Lukas, Claudia Engel, and Camilla Mazzucato, "Towards a Living Archive: Making Multi Layered Research Data and Knowledge Generation Transparent," *Journal of Field Archaeology* 43, no. 1 (2018): S19–S30, https:// www.tandfonline.com/doi/full/10.1080/00934690.2018.1516110.

81. Jeremy Pilcher and Saskia Vermeylen, "From Loss of Objects to Recovery of Meanings: Online Museums and Indigenous Cultural Heritage," *M/C Journal* 11, no. 6 (2008): 3, https://doi.org/10.5204/mcj.94.

82. Karen Ingersoll, *Waves of Knowing: A Seascape Epistemology* (Durham: Duke University Press, 2016).

83. Julie Michelle Klinger, "Lithosociality in and in Relation to Outer Space" (paper presented at Society of Social Studies of Science Conference, Virtual Meeting, October 6, 2021).

84. Stephanie LeMenager, *Veer Ecology: A Companion for Environmental Thinking*, ed. Jeffrey Jerome Cohen and Lowell Duckert (Minneapolis: University of Minnesota Press, 2017), locations 3979–80 of 12937, Kindle.

85. LeMenager, locations 3962–64 of 12937, Kindle.

86. Barron, "Drowned in Law," 160.

87. Pearl Harbor Historic Sites, "USS *Arizona* Memorial," last updated 2021, https://www.pearlharborhistoricsites.org/pearl-harbor/arizona-memorial.

88. Philip Turner et al., "Memorializing the Middle Passage on the Atlantic Seabed in Areas beyond National Jurisdiction," *Marine Policy* 122 (2020): 104254, https://doi.org/10.1016/j.marpol.2020.104254.

89. Kathryn Yusoff, "Geologic Life: Prehistory, Climate, Futures in the Anthropocene," *Environment and Planning D: Society and Space* 31, no. 5 (2013): 782.

90. Epeli Hau'ofa, "Our Sea of Islands," in *A New Oceania: Rediscovering Our Sea of Islands*, ed. Eric Waddell, Vijay Naidu, and Epeli Hau'ofa (Suva, Fiji: University of the South Pacific, 1993), 8.

91. Melody Jue, "Anthropocene Chemistry: Residual Media after Deep water Horizon," talk given at UCSB, December 5, 2018, http://ejcj.orfaleacen ter.ucsb.edu/2018/11/7069/.

92. See Shreema Mehta, "The Dangers of Deep Sea Mining," *Earthworks*, September 28, 2015, https://www.earthworksaction.org/earthblog/detail/the_dangers_of_deep_sea_mining#.VuNyUJMrKb8; and Kathryn A. Miller et al., "An Overview of Seabed Mining Including the Current State of Development, Environmental Impacts, and Knowledge Gaps," *Frontiers in Marine Science* 4 (2018): 418.

93. This is a riff on what Joshua Scannell termed "deep managerial time," as neoliberalist "ontological stabilization of populations." Joshua Scannell, "Both a Cyborg and a Goddess: Deep Managerial Time and Informatic Governance," in *Object Oriented Feminism*, ed. Katherine Behar (Minneapolis: University of Minnesota Press, 2016), 5.

2. Swimmers in a Sonic Pipeline

1. potoutos, "Hearing Seismic Surveys While Underwater," YouTube, 1:20, December 9, 2013, https://youtu.be/aoXhmsPXlLI.

2. ge0physicsrocks, "3D Seismic," YouTube, 4:27, August 29, 2011, https://www.youtube.com/watch?v=hxJa7EvYoFI.

3. Frances Dyson, *The Tone of Our Times: Sound, Sense, Economy, and Ecology* (Cambridge: MIT Press, 2014), 52.

4. J. Martin Daughtry, "Thanatosonics: Ontologies of Acoustic Violence," *SocialText* 32, no. 2 (2014): 36. Daughtry categorizes certain noises as "belliphonic" sound, "the vehicular, weapon-related, and other sounds that armed combat produces." See J. Martin Daughtry, *Listening to War: Sound, Music, Trauma, and Survival in Wartime Iraq* (Oxford: Oxford University Press, 2015), 5, 33.

5. Writing about the early development of sonar, John Shiga delineates a semiotic taxonomy of pings and echoes specific to underwater transmission and perception. John Shiga, "Sonar: Empire, Media, and the Politics of Underwater Sound," *Canadian Journal of Communication* 38 (2013): 357–77.

6. To hear a sample of a seismic airgun survey, see "Seismic Airgun Surveys," *Ocean Conservation Research*, http://ocr.org/portfolio/seismic-airgun-sur veys/.

7. Alfred Keil, "The Development of Ocean-Engineering (An Attempt at a Methodology)," Department of Naval Architecture and Marine Engineering (September 8, 1966), 8, MIT Oral History Program, oral history interviews on

ocean engineering, MC-0089, box 4, "Transportation by Sea—Today and Tomorrow," by A. A. H. Keil, Department of Distinctive Collections, MIT Libraries, Cambridge, Massachusetts.

8. ICOEES Typoe Task Group, Executive Summary of Preliminary Position Paper, "An International Decade of Ocean Exploration and Assessment of the Seas," Committee on Ocean Engineering, January 12, 1968, Rand (William W.) Papers ca. 1921–1968, SBHC Mss 46, box 1, Department of Special Collections, Davidson Library, University of California, Santa Barbara.

9. *Merriam-Webster*, s.v. "survey," https://www.merriam-webster.com/dictionary/survey.

10. William Whitehall Rand, "Santa Barbara Channel, Offshore Oil Exploration," to Rotary Club, 1957, p. 2, Rand Papers, SBHC Mss 46, box 1.

11. Andrea Ballestero, *A Future History of Water* (Durham: Duke University Press, 2019), 15.

12. Stephanie LeMenager, *Living Oil: Petroleum Culture in the American Century* (Oxford: Oxford University Press, 2014); Darin Barney, "Pipelines," in *Fueling Culture: Politics, History, Energy*, ed. Imre Szeman, Jennifer Wenzel, and Patsy Yaeger (New York: Fordham University Press, 2017), 267–70; Imre Szeman, "System Failure: Oil, Futurity, and the Anticipation of Disaster," *South Atlantic Quarterly* 106, no. 4 (2007): 805–23.

13. Nicole Starosielski, "Beyond Fluidity: A Cultural History of Cinema under Water," in *Ecocinema Theory and Practice*, ed. Stephen Rust, Salma Monani, and Sean Cubitt (New York: Routledge, 2012), 157.

14. *Oxford English Dictionary*, s.v. "sound," https://www.oed.com/view/Entry/185130?isAdvanced=false&result=9&rskey=7deW68&.

15. Stefan Helmreich, *Sounding the Limits of Life: Essays in the Anthropology of Biology and Beyond* (Princeton: Princeton University Press, 2016), 185.

16. Stefan Helmreich, *Alien Ocean: Anthropological Voyages in Microbial Seas* (Berkeley: University of California Press, 2009), 34.

17. Gary Weir, *An Ocean in Common: American Naval Officers, Scientists, and the Ocean Environment* (College Station: Texas A&M University Press, 2001), 6.

18. Shiga, "Sonar."

19. "Shannon and Weaver Model of Communication," *Communication Theory*, https://www.communicationtheory.org/shannon-and-weaver-model-of-communication/.

20. Shiga, "Sonar," 362.

21. Shiga, 365.

22. "The First Practical Uses of Underwater Acoustics: The Early 1900s," *Discovery of Sound in the Sea*, https://dosits.org.

23. Jeffrey A. Karson et al., *Discovering the Deep: A Photographic Atlas of the Seafloor and Ocean Crust* (Cambridge, UK: Cambridge University Press, 2015), 4.

24. Weir, *An Ocean in Common*, 30.

25. Weir, 14.

26. Submarine Signal Company, "The Development of the Fathometer and Echo Depth Finding," *Soundings*, April 11, 1932.

27. Ronald Rainger, "Science at the Crossroads: The Navy, Bikini Atoll, and American Oceanography in the 1940's," *Historical Studies in the Physical and Biological Sciences* 30, no. 2 (2000): 352–53, reprinted courtesy of the History of the Sciences Society from Earth Sciences History, 2000.

28. Henry M. Stommel, "Columbus O'Donnell Iselin," in *Biographical Memoirs*, vol. 64 (Washington, D.C.: National Academy Press, 1994), https://www.nap.edu/read/4547/chapter/8.

29. Shiga, "Sonar," 367–68.

30. Columbus O'D. Iselin, "Some Phases of Modern Deep-Sea Oceanography," in *Annual Report of the Board of Regents of the Smithsonian Institution* (Washington, D.C.: United States Government Printing Office, 1932), 258, http://library.si.edu/digital-library/book/annualreportofbo1932smit.

31. "Entanglement of Marine Life: Risks and Response," NOAA Fisheries, June 19, 2017, https://www.fisheries.noaa.gov/insight/entanglement-marine-life-risks-and-response.

32. Rand, "Santa Barbara Channel," 1.

33. Rand, 3.

34. E. C. LaFond, Robert S. Dietz, and J. A. Knauss, "A Sonic Device for Underwater Sediment Surveys," U.S. Navy Electronics Laboratory, Oceanographic Studies Section, San Diego 52, California, *Journal of Sedimentary Petrology* 20, no. 2 (June 1950): 108.

35. LaFond, Dietz, and Knauss, 110.

36. Don Ihde, *Listening and Voice: Phenomenologies of Sound* (New York: SUNY Press, 2007), 67.

37. Bill Dragoset "A Historical Reflection on Reflections," *The Leading Edge* 24 (2005): S48, https://doi.org/10.1190/1.2112392.

38. Nicole Starosielski, *The Undersea Network* (Durham: Duke University Press, 2015), location 539 of 9506, Kindle.

39. Huidong Li, Z. Daniel Deng, and Thomas J. Carlson, "Piezoelectric Materials Used in Underwater Acoustic Transducers," *Sensor Letters* 10, nos. 3–4 (2012): 679–97, http://jsats.pnnl.gov/Publications/Peer/2012/2012_Li_etal_PZT_Review_paper_Sensor_Letters.pdf.

40. Dragoset, "A Historical Reflection," S46.

41. Ibrahim Palaz and K. J. Marfurt, eds., *Carbonate Seismology* (Tulsa: Society of Exploration Geophysicists, 1997), 40.

42. Eric Roberts, "Detonation and Combustion," Stanford, https://cs.stanford.edu/people/eroberts/courses/ww2/projects/firebombing/detonation-and-combustion.htm.

43. David M. Lawrence, *Upheaval from the Abyss: Ocean Floor Mapping and the Earth Science Revolution* (New Brunswick, N.J.: Rutgers University Press, 2002), location 1535 of 3147, Kindle.

44. Herbert Grove Dorsey, "The Transmission of Sound through Sea Water. II," *Journal of the Acoustical Society of America* 7 (1936): 287–99.

45. John Jakosky, "Characteristics of Explosives for Marine Seismic Exploration," *Geophysics* 21, no. 4 (1956): 969–91, https://doi.org/10.1190/1.1438316.

46. Correspondence to Dr. Robert A. Forsch, Hudson Laboratories, from Bob Westervelt, August 5, 1957, USNUSL, John Brackett Hersey Papers, MC-12, box 9, folder 2, Hersey, J. Brackett Correspondence—U.S. Navy Underwater Sound Laboratory, 1948–1958, Woods Hole Oceanographic Institution, Data Library and Archives (hereafter WHOI), Woods Hole, Massachusetts, http://archives.mblwhoilibrary.org:8081/repositories/2/resources/17.

47. Correspondence to J. B. Hersey from B. J. O'Keefe (EGG), November 24, 1960, 1, John Brackett Hersey Papers, MC-12, box 9, folder 2, Hersey, J. Brackett Correspondence—Edgerton, Germeshausen & Grier, Inc., 1960, WHOI.

48. The Geological Society of America, "Memorial to John Brackett Hersey, 1913–1992," 207–9, http://www.geosociety.org/documents/gsa/memorials/v24/Hersey-JB.pdf.

49. O. Leenhardt, "Analysis of Continuous Seismic Profiles," *International Hydrographic Review* 46, no. 1 (2015), https://journals.lib.unb.ca/index.php/ihr/article/view/23959.

50. J. B. Hersey, "Sound Reflections in and under Oceans," reprinted from *Physics Today*, November 1965, 17–24, John Brackett Hersey Papers, MC-12, box 16, folder 2, Articles, abstracts reprints 1944–1982, WHOI.

51. Harold E. Edgerton, "The 'Boomer' Sonar Source for Seismic Profiling," *Journal of Geophysical Research* 69, no. 8 (April 15, 1964): 3033–42, Hersey Papers, MC-12, box 5, folder 2, H. Edgerton's paper, The Boomer Sonar Source for Seismic Profiling, WHOI.

52. Hersey, "Sound Reflections," 23.

53. S. T. Knott and E. T. Bunce, "Recent Improvement in Technique of Continuous Seismic Profiling," *Deep Sea Research* 15, no. 5 (1968): 638, https://doi.org/10.1016/0011-7471(68)90072-7.

54. Dragoset, "A Historical Reflection," S54.

55. Dragoset, S54.

56. "Seismic Surveys," *Beachapedia*, http://www.beachapedia.org/Seismic_Surveys.

57. John Brackett Hersey, "Speech Delivered in Annapolis June 23, 1971," pp. 9–12, Hersey Papers, MC-12, box 16, folder 1, Hersey, J. Brackett Speeches, 1955–1980, WHOI.

58. "IPOD Site Survey Criteria: Multichannel Seismic Surveys of IPOD Sites," IPOD Site Survey Guidelines, November 1975, p. 23. Deep Earth Sampling Executive and Planning Committees, 1975–1976, College of Oceanic and Atmospheric Sciences Records, RG 173, Series II, Box 1, Oregon State University Special Collections and Archives Research Center, Corvallis, Oregon.

59. Dragoset, "A Historical Reflection," S67.

60. Donald E. Koelsch et al., "A Deep Towed Explosive Source for Seismic Experiments on the Ocean Floor," *Marine Geophysical Research* 8 (1986): 345–46, https://doi.org/10.1007/BF02084018.

61. Frank Tursi, "A Very Brief History of Offshore Drilling," *Coastal Review Online*, https://www.coastalreview.org/2015/06/a-very-brief-history-of-offshore-drilling/.

62. James Broda, interview with author, Woods Hole Oceanographic Institution, June 25, 2018.

63. On a related note, Susan Douglas speaks of "dimensional listening," a term that she coined to describe the way in which 1920s radio shows prompted listeners to construct spatial imaginaries of three-dimensional locales, like a ballpark or a cityscape. Susan J. Douglas, *Listening In: Radio and the American Imagination* (Minneapolis: University of Minnesota Press, 1999), 33.

64. According to the Institute for Energy Research, "The Outer Continental Shelf (OCS) is the submerged area between a continent and the deep ocean. It is a rich natural resource for the deep ocean. It is a rich natural resource for the United States, containing an estimated 86 billion barrels of oil and 420 trillion cubic feet of natural gas." "Outer Continental Shelf," Institute for Energy Research, http://instituteforenergyresearch.org/topics/policy/ocs/.

65. "Offshore Access to Oil and Natural Gas Resources," February 2015, American Petroleum Institute report, 1, 7, http://www.api.org/~/media/files/oil-and-natural-gas/offshore/offshoreaccess-primer-lores.pdf.

66. According to Sterne, through stethoscopes, mute bodies became sounding ones, shifting the locus of truth away from what patients say and toward what bodies reveal. Jonathan Sterne, *The Audible Past: Cultural Origins of Sound Reproduction* (Durham: Duke University Press, 2003), 117.

67. Veit Erlmann, *Reason and Resonance: A History of Modern Aurality* (New York: Zone Books, 2014), 151.

68. *Online Etymology Dictionary*, s.v. "noise," https://www.etymonline.com/word/noise.

69. Helmreich, *Sounding the Limits of Life*, xix.

70. Emily Thompson, for instance, talks about historical periods in which sonic culture is defined by noisy din, while Brian Larkin discusses the differential and socially layered filtering of noise and signal within urban soundscapes in Nigeria. Emily Thompson, *The Soundscape of Modernity: Architectural Acoustics and the Culture of Listening in America, 1900–1933* (Cambridge: MIT Press, 2002); Brian Larkin, *Signal and Noise: Media, Infrastructure, and Urban Culture in Nigeria* (Durham: Duke University Press, 2008).

71. Michel Serres, *The Parasite*, trans. Lawrence R. Schehr (Baltimore: Johns Hopkins University Press, 1982), 13.

72. "Rice's Whale," *Defenders of Wildlife*, January 2021, https://defenders.org/wildlife/rices-whale.

73. "North Atlantic Right Whale," updated December 12, 2023, NOAA Fisheries, https://www.fisheries.noaa.gov/species/north-atlantic-right-whale.

74. John A. Hildebrand, "Anthropogenic and Natural Sources of Ambient Noise in the Ocean." *Marine Ecology Progress Series* 395 (2009): 5–20.

75. Leila T. Hatch et al., "Quantifying Loss of Acoustic Communication Space for Right Whales in and around a U.S. National Marine Sanctuary," *Conservation Biology* 26 (2012): 983–94, https://doi.org/10.1111/j.1523-1739.2012.01908.x.

76. Margret Grebowicz, *Whale Song (Object Lessons)* (New York: Bloomsbury Publishing, 2017), location 107 of 2251, Kindle.

77. "A Noisy Ocean: A Q&A with Dr. Leila Hatch," NOAA National Marine Sanctuaries, June 2016, https://sanctuaries.noaa.gov/news/jun16/noisy-ocean.html.

78. Cascadia Research Collective, "Summer Feeding Areas, Winter Feeding Areas and Migration," http://www.cascadiaresearch.org/splash-structure-populations-levels-abundance-and-status-humpback-whales-north-pacific/summer.

79. Maya Yamato and Nicholas D. Pyenson, "Early Development and Orientation of the Acoustic Funnel Provides Insight into the Evolution of Sound Reception Pathways in Cetaceans," *PLOS ONE* 10, no. 3 (2015): 1–15, https://doi.org/10.1371/journal.pone.0118582.

80. Yamato and Pyenson.

81. Gilles Deleuze and Félix Guattari, "The Eye and the Hand," in *The Logic of Sense*, trans. Mark Lester (New York: Columbia University Press, 1990).

82. See Colin Milburn, *Nanovision: Engineering the Future* (Durham: Duke University Press, 2008), 85.

83. Douglas, *Listening In*, 33–34.

84. Steve Goodman, *Sonic Warfare: Sound, Affect, and the Ecology of Fear* (Cambridge: MIT Press, 2010), 9.

85. See Raviv Ganchrow, "Earth-Bound Sound: Oscillations of Hearing, Ocean, and Air," *Theory & Event* 24, no. 1 (2021): 67–116.

86. Michael Gallagher, Anja Kanngieser, and Jonathan Prior, "Listening Geographies: Landscape, Affect, and Geotechnologies," *Progress in Human Geography* 41, no. 5 (2017): 630, http://dx.doi.org/10.1177/0309132516652952.

87. Alfred North Whitehead, *Process and Reality: An Essay in Cosmology* (New York: The Free Press, 1978), 29; Jennifer Gabrys, *Program Earth: Environmental Sensing Technology and the Making of a Computational Planet* (Minneapolis: University of Minnesota Press, 2016), 13.

88. Joshua Horowitz, *War of the Worlds: A True Story* (New York: Simon & Schuster, 2015).

89. "Navy Agrees to Limit Global Sonar Deployment," *Natural Resources Defense Council* Press Release, October 13, 2003, https://www.nrdc.org/media/2003/031013.

90. Donald L. Evans and Gordon R. England, "Joint Interim Report: Bahamas Marine Mammal Stranding Event of 15–16 March 2000," U.S. Department of Commerce, Secretary of the Navy, National Oceanic and Atmospheric Administration, National Marine Fisheries Service, December 2001, http://www.nmfs.noaa.gov/pr/pdfs/health/stranding_bahamas2000.pdf.

91. Andrew Darby, "Navy Rejects Whale Blame," *theage.com.au*, October 27, 2005. http://www.theage.com.au/news/national/navy-rejects-whale-blame/2005/10/26/1130302839888.html.

92. Evans and England, "Joint Interim Report," iii.

93. Evans and England.

94. Elizabeth Burgess, interview with author, New England Aquarium, March 25, 2022.

95. Arne Kalland, "Management by Totemization: Whale Symbolism and the Anti-whaling Campaign," *Arctic* 46, no. 2 (June 1993): 124–33.

96. Several animal studies authors have explored the alignments between racism and speciesism. See Cary Wolfe, *Animal Rites: American Culture, the Discourse of Species, and Posthumanist Theory* (Chicago: University of Chicago Press, 2008); Claire Jean Kim, *Dangerous Crossings: Race, Species, and Nature in a Multicultural Age* (Cambridge, UK: Cambridge University Press, 2015); Zakiyyah Iman Jackson, *Becoming Human: Matter and Meaning in an Antiblack World* (New York: New York University Press, 2020).

97. Resolution of the Board of Supervisors of the County of Santa Barbara, #8242, passed July 19, 1948.

98. Resolution of the Board.

99. *Marine Seismograph Survey for Union Oil Company of California of 80&81 Prospects* (Pasadena: United Geophysical Co., Inc., 1948), 14.

100. Fred Aminzadeh and Shivaji N. Dasgupta, *Geophysics for Petroleum Engineers* (Amsterdam: Elsevier, 2013), 54.

101. Lindy Weilgart, "A Review of the Impacts of Seismic Airgun Surveys on Marine Life," submitted to the CBD Expert Workshop on Underwater Noise and Its Impacts on Marine and Coastal Biodiversity, February 25–27, 2014, London, UK, 2013, https://www.cbd.int/doc/?meeting=MCBEM-2014 -01. Also see Conservation and Development Problem Solving Team, "Anthropogenic Noise in the Marine Environment," prepared for the National Oceanic and Atmospheric Administration and the Marine Conservation Biology Institute, December 5, 2000, http://sanctuaries.noaa.gov/management/pdfs/anthro_noise.pdf; and "Boom, Baby, Boom: The Environmental Impacts of Seismic Surveys," *National Resources Defense Council*, May 2010, https://www.nrdc.org/oceans/files/seismic.pdf.

102. Thomas Keevin and Gregory Hempen, "Environmental Effects of Underwater Explosions with Methods to Mitigate Impacts" (St. Louis: U.S. Army Corps of Engineers, 1997), 23.

103. Robert McCauley et al., "Widely Used Marine Seismic Survey Air Gun Operations Negatively Impact Zooplankton," *Nature Ecology & Evolution* 1, no. 0195 (2017): 1, https://www.nature.com/articles/s41559-017-0195.

104. Amy Propen, *Locating Visual-Material Rhetorics: The Map, the Mill, and the GPS* (Anderson, S.C.: Parlor Press, 2012), 165–67.

105. Propen, 179.

106. Kraus qtd. in Craig Lemoult, "Researchers Worry Right Whales Could Be Harmed during Seismic Testing," *NPR*, April 14, 2019, https://www.npr.org/2019/04/15/713387959/researchers-worry-right-whales-could-be -harmed-during-seismic-testing.

107. Burgess interview.

108. Joanne O'Brien, Simon Berrow, and Dave Wall, "The Impact of Multibeam on Cetaceans: A Review of Best Practice," Irish Whale and Dolphin Group,

March 2005, 5, http://www.ecomarbelize.org/uploads/9/6/7/0/9670208/multi beam__1_.pdf.

109. Keevin and Hempen, "Environmental Effects of Underwater Explosions," 74.

110. Lucia Di Iorio and Christopher W. Clark, "Exposure to Seismic Survey Alters Blue Whale Acoustic Communication," *Biology Letters* 6, no. 1 (2010): 51–54, https://doi.org/10.1098/rsbl.2009.0651.

111. Chao Peng, Zhao Xinguo, and Guangxu Liu, "Noise in the Sea and Its Impacts on Marine Organisms," *International Journal of Environmental Research and Public Health* 12, no. 10 (2015): 12304–23, https://doi.org/10.3390/ijerph 121012304.

112. "Incidental Take Authorizations for Military Readiness Activities," NOAA Fisheries, https://www.fisheries.noaa.gov/national/marine-mammal-pro tection/incidental-take-authorizations-military-readiness-activities.

113. National Resources Defense Council, "In a Blow to Marine Life, Trump Administration Greenlights Seismic Blasting in Atlantic," *NRDC Expert Blog*, November 30, 2018, https://www.nrdc.org/experts/nrdc/blow-marine-life -trump-administration-greenlights-seismic-blasting-atlantic?fbclid=IwAR2Vhj FA1GbXoi5IoGAeqahCoZJ3wXI1U4LgL7pgRJRw8Uou8lo0xadA29c.

114. American Petroleum Institute, "Offshore Seismic Surveys: Why and How," 2021, https://www.api.org/oil-and-natural-gas/energy-primers/offshore/ offshore-seismic-surveys-why-how.

115. Ned Rossiter defines *logistical media* as "technologies, infrastructure, and software" that "coordinate, capture, and control the movement of people, finance, and things." Ned Rossiter, *Software, Infrastructure, Labor: A Media Theory of Logistical Nightmares* (New York: Routledge, 2016), locations 4–5 of 6203, Kindle.

116. See Carole Stabile, "Shooting the Mother: Fetal Photography and the Politics of Disappearance," *Camera Obscura* 28 (1992): 180; Rosalind Pollack Petchesky, "Fetal Images: The Power of Visual Culture in the Politics of Reproduction," *Feminist Studies* 13, no. 2 (1987): 265; Valerie Hartouni, *Cultural Conceptions: On Reproductive Technologies and the Remaking of Life* (Minneapolis: University of Minnesota Press, 1997).

117. American Petroleum Institute, "The Offshore Energy We Need," https://www.api.org/oil-and-natural-gas/energy-primers/offshore/the-offshore -energy-we-need.

118. Petroleum Exploration and Production Association of New Zealand, "Seismic Surveys: Exploring What Lies Beneath," http://www.seismicsurvey .co.nz/.

119. Rossiter, *Software, Infrastructure, Labor*, 10.

120. Szeman, "System Failure," 812.

121. Achille Mbembe, "Necropolitics," trans. Libby Meintjes, *Public Culture* 15, no. 1 (2003): 27.

122. See Kathryn A. Miller et al., "An Overview of Seabed Mining Including the Current State of Development, Environmental Impacts, and Knowledge

Gaps," *Frontiers in Marine Science* 4 (2018): 418, https://doi.org/10.3389/fmars
.2017.00418; Deep Ocean Stewardship Initiative, "Deep Ocean Stewardship Initiative: Advancing Science-Based Policy," http://dosi-project.org/.

123. Jonathan Sterne, "Sonic Imaginations," in *The Sound Studies Reader*, ed. Jonathan Sterne (New York: Routledge, 2012), 9.

124. Tim Ingold, *The Perception of the Environment: Essays on Livelihood, Dwelling, and Skill* (New York: Routledge, 2002), 7.

125. The word *anneal* is typically used to describe a process of combining substances—typically glass, steel, or DNA—through a heating and cooling process, that permanently changes the original substances in their mixing, resulting in a tougher and stronger product. My choice in using this term is meant to evoke this sense of a strengthening, essential change that is constituted by the annealing of our oceanic images to the saturated bodies and materials that produce it.

3. Perilous Plumes

1. "Hydrothermal Vents," New Millennium Observatory Explorer, https://www.pmel.noaa.gov/eoi/nemo/explorer/concepts/hydrothermal.html.

2. Unlike photosynthesis, chemosynthesis produces energy through oxidation reactions with inorganic compounds like hydrogen sulfide.

3. Jillian Peterson et al., "Dual Symbiosis of the Vent Shrimp *Rimicaris exoculata* with Filamentous Gamma- and Epsilonproteobacteria at Four Mid-Atlantic Ridge Hydrothermal Vent Fields," *Environmental Microbiology* 12, no. 8 (2010): 2205, https://onlinelibrary.wiley.com/doi/pdf/10.1111/j.1462-2920.2009.02129.x; Joel W. Martin and Timothy M. Shank, "A New Species of the Shrimp Genus *Chorocaris* (Decapoda: Caridea: Alvinocarididae) from Hydrothermal Vents in the Eastern Pacific Ocean," *Proceedings of the Biological Society of Washington* 118, no. 1 (April 2005): 196.

4. "Subsea Mining—Deep Sea Ocean Mining and Seafloor Dredging Operations," EddyPump Corporation, https://eddypump.com/education/sub sea-mining-deep-sea-dredging/.

5. National Oceanic and Atmospheric Administration, "Deep-Sea Mining Interests in the Clarion-Clipperton Zone," NOAA Ocean Exploration, https://oceanexplorer.noaa.gov/explorations/18ccz/background/mining/mining.html.

6. Leonard Engel, "Exploring the Great Deeps," *Harper's Magazine*, February 1960, 48, Elizabeth T. "Betty" Bunce Papers, MC 21, box 1, Woods Hole Oceanographic Institution, Data Library and Archives, Woods Hole, Massachusetts.

7. Roopali Phadke, "Green Energy Futures: Responsible Mining on Minnesota's Iron Range," *Energy Research & Social Science* 35 (2018): 163–73.

8. John L. Mero, "Minerals on the Ocean Floor," *Scientific American* 203, no. 6 (December 1960): 64–73, https://www.jstor.org/stable/24940721.

9. Wil Hyton, "History's Largest Mining Operation Is about to Begin," *Atlantic*, January/February 2020, https://www.theatlantic.com/magazine/archive/2020/01/20000-feet-under-the-sea/603040/.

10. Rachel Reeves, "My Family's Pacific Island Home Is Grappling with Deep-Sea Mining," *Hakai Magazine*, November 30, 2021, https://hakaimaga zine.com/features/my-familys-pacific-island-home-is-grappling-with-deep -sea-mining/?fbclid=IwAR2rv5Uvxmn_xmqQ0WPE_h36vKUzzgbHk6OP k5Y7jfjlHk88pkWORl_EuWU.

11. Rob Nixon, *Slow Violence and the Environmentalism of the Poor* (Cambridge, Mass.: Harvard University Press, 2011).

12. See Melody Jue, "Proteus and the Digital: Scalar Transformations of Seawater's Materiality in Ocean Animations," *Animation* 9, no. 2 (2014): 246.

13. *Oxford English Dictionary*, s.v. "turbulent," http://www.oed.com/view/ Entry/207572?redirectedFrom=turbulent#eid.

14. S. A. Thorpe, *The Turbulent Ocean* (Cambridge, UK: Cambridge University Press, 2005), 3.

15. Helmreich defines mediation ecologically as the "propagation of action through substances, channels, or instruments." Stefan Helmreich, *Alien Ocean: Anthropological Voyages in Microbial Seas* (Oakland: University of California Press, 2009), 32.

16. Jeffrey Cohen, "Elemental Relations," *O-Zone: A Journal of Object-Oriented Studies* 1 (2014): 58.

17. Andrea Ballestero, "The Plume: Movement and Mixture in Subterranean Water Worlds," *Fieldsights*, September 22, 2020, https://culanth.org/field sights/the-plume-movement-and-mixture-in-subterranean-water-worlds.

18. Elizabeth A. Povinelli, "Fires, Fogs, Winds," *Cultural Anthropology* 32, no. 4 (2017): 508, https://doi.org/10.14506/ca32.4.03.

19. Kirsten Ostherr calls this "inoculation by representation." Ostherr, *Cinematic Prophylaxis: Globalization and Contagion in the Discourse of World Health* (Durham: Duke University Press, 2005), 2.

20. Cohen, "Elemental Relations," 55.

21. G. R. Hunt and T. S. Van Den Bremer, "Classical Plume Theory: 1937–2010 and Beyond," *MA Journal of Applied Mathematics* 76 (2011): 425; Eckart Meiburg, "Modeling the Pacific Ocean on the Computer," Carsey-Wolf Center, March 22, 2021, https://vimeo.com/527398493.

22. Lisa Cartwright, *Screening the Body: Tracing Medicine's Visual Culture* (Minneapolis: University of Minnesota Press, 1995), 20.

23. Jonathan Crary, *Techniques of the Observer: On Vision and Modernity in the Nineteenth Century* (Cambridge: MIT Press, 1993), 3.

24. Jussi Parikka, *A Geology of Media* (Minneapolis: University of Minnesota Press, 2015), 1.

25. Parikka, 14; Mary Beth Gallagher, "Understanding the Impact of Deep-Sea Mining," *MIT News*, December 5, 2019, http://news.mit.edu/2019/ understanding-impact-deep-sea-mining-1206.

26. Stefanie Hessler, *Prospecting Ocean* (Cambridge: MIT Press, 2019), 167.

27. Blue Nodules, "Blue Nodules Summary—Deep Sea Mining," YouTube, 7:34, August 3, 2020, https://youtu.be/pCus0hTsibc.

28. Erich Auerbach, "Figura," in *Scenes from the Drama of European Literature* (Manchester: Manchester University Press, 1984), 49, qtd. in Juan Llamas-Rodriguez, "Narcontologies: Toward a Media Ontological Approach to Narcotrafficking" (PhD diss., University of California, Santa Barbara, 2017), 13.

29. The name "MTT" is derived from the 1956 Morton, Taylor, and Turner study which developed seminal theories of turbulent gravitational convection from maintained and instantaneous sources of buoyancy. Hunt and Van Den Bremer, "Classical Plume Theory," 424.

30. F. A. Gifford, "Atmospheric Dispersion Models for Environmental Pollution Applications," in *Lectures on Air Pollution and Environmental Impact Analyses* (Boston: American Meteorological Society, 1982), 35, https://doi.org/10.1007/978-1-935704-23-2_2.

31. Denis Cosgrove, "Contested Global Visions: One-World, Whole-Earth, and the Apollo Space Photographs," *Annals of the Association of American Geographers* 84, no. 2 (1994): 270–94.

32. J. E. Lupton et al., "Entrainment and Vertical Transport of Deep-Ocean Water by Buoyant Hydrothermal Plumes," *Nature* 316, no. 15 (August 1985): 621–23; Sarah A. Little, Keith D. Stolzenbach, and Richard P. Von Herzen, "Measurements of Plume Flow from a Hydrothermal Vent Field," *Journal of Geophysical Research: Solid Earth* 92, no. B3 (1987): 2587–96.

33. Ecological or systems-oriented perspectives on the planet emerged in the 1970s with thinkers such as Marshall McLuhan, Steward Brand, and Buckminster Fuller, who saw connections between the rise of information communications technology and global, "Whole Earth" perspectives. The Earth itself was increasingly thought of as a singular, adaptive control system—a view that was further established through satellite photography and the internet. See Cosgrove, "Contested Global Visions."

34. Jenette Restivo and Daniel Cojanu, "Ch. 4: Searching for the Plume," in *Science in a Time of Crisis: WHOI's Response to the Deepwater Horizon Oil Spill* (Woods Hole, Mass.: Woods Hole Oceanographic Institution, 2011), https://www.whoi.edu/deepwaterhorizon/chapter4.html.

35. Lonny Lippsett, "A Plume of Chemicals from Deepwater Horizon," *Oceanus*, Woods Hole Oceanographic Institution, July 18, 2011, https://www.whoi.edu/oceanus/feature/a-plume-of-chemicals-from-deepwater-horizon/.

36. Lippsett.

37. See Meiburg, "Modeling the Pacific Ocean."

38. Melody Jue, "Fluid Cuts: The Anti-visual Logic of Surfactants after *Deepwater Horizon*," *Configurations* 27, no. 4 (2019): 528.

39. Kesava Reddy and Mark D'Andrea, "Health Consequences among Subjects Involved in Gulf Oil Spill Clean-up Activities," *American Journal of Medicine* 26, no. 11 (November 2013): 966–74, https://doi.org/10.1016/j.amjmed.2013.05.014.

40. Rodrigo Almeda, Cammie Hyatt, and Edward J. Buskey, "Toxicity of Dispersant Corexit 9500A and Crude Oil to Marine Microzooplankton," *Ecotoxicology and Environmental Safety* 106 (August 2014): 76–85.

41. Carlos Muñoz Royo, Zoom interview with author, October 14, 2021.

42. Daniel O. B. Jones et al., "Biological Responses to Disturbance from Simulated Deep-Sea Polymetallic Nodule Mining," *PLOS ONE* 12, no. 2 (2017): e0171750, https://doi.org/10.1371/journal.pone.0171750.

43. R. E. Burns, "Assessment of Environmental Effects of Deep Ocean Mining of Manganese Nodules," DOMES Project/NOAA, *Helgoländer Meeresunters* 33 (1980): 436.

44. Jones et al., "Biological Responses to Disturbance."

45. Laura Martin offers extensive background on the relationship between Cold War nuclear science and environmental science, which notably includes deep-ocean research. Specifically, methods of measuring radioactivity in the aftermath of nuclear violence enabled techniques like radioautographs and radioactive tracers, which could be used to image animal interiors, discover trophic relationships between species, and, in this case, track the movements of matter in fluid. Indeed, Martin contends that nuclear violence and the radioactive research that followed contributed to the very concept of the ecosystem: "Violence made ecosystems manifest" (569). Laura Martin, "Proving Grounds: Ecological Fieldwork in the Pacific and the Materialization of Ecosystems," *Environmental History* 23 (2018): 567–92.

46. Gary Taghon, "Determining Effects of Manganese Nodule Mining on Deep-Sea Benthic Communities: Planning and Evaluating a Controlled Impact Statement," National Oceanic and Atmospheric Administration to Gary L. Taghon, OSU College of Oceanography, September 10, 1986, box 37, Research Accounting Office Records (RG 26), Oregon State University Archives, Corvallis, Oregon.

47. Massachusetts Institute of Technology, "Mining the Deep Sea," YouTube video, December 5, 2019, https://youtu.be/MWvCtF1itQM.

48. Royo interview.

49. Thomas Peacock and Matthew H. Alford, "Is Deep-Sea Mining Worth It?," *Scientific American* 318, no. 5 (May 2018): 76, https://www.scientificameri can.com/index.cfm/_api/render/file/?method=inline&fileID=386AD 132-4C70-4CC9-BDAF0455741F89A1.

50. See Adrienne Buller, *The Value of a Whale: On the Illusions of Green Capitalism* (Manchester: Manchester University Press, 2022).

51. Cindy Dover, phone interview with author, January 16, 2018.

52. R. E. Burns et al., "Observations and Measurements during the Monitoring of Deep Ocean Manganese Nodule Mining Test in the North Pacific, March–May 1978," Boulder, Colorado, U.S. Department of Commerce, NOAA Technical Memorandum, ERL MESA, https://play.google.com/books/reader?id =3XkeAQAAIAAJ&pg=GBS.PR4&hl=en.

53. Royo interview.

54. Christopher W. Pinto quoted in Alina Jaeckel, *The International Seabed Authority and the Precautionary Principle* (Boston: Brill Nijhoff, 2017), 5.

55. Bishnupriya Ghosh and Bhaskar Sarkar, "Media and Risk: An Introduction," in *The Routledge Companion to Media and Risk*, ed. Bishnupriya Ghosh and Bhaskar Sarkar (New York: Routledge, 2020), 3.

56. Aimee Bahng, *Migrant Futures: Decolonizing Speculation in Financial Times* (Durham: Duke University Press, 2017), 5.

57. Molly Wallace, *Risk Criticism: Precautionary Reading in an Age of Environmental Uncertainty* (Ann Arbor: University of Michigan Press, 2016), 20.

58. International Seabed Authority, *Developing a Regulatory Framework for Mineral Exploitation in the Area: A Discussion Paper on the Development and Drafting of Regulations on Exploitations for Mineral Resources in the Area (Environmental Matters)* (Kingston, Jamaica: International Seabed Authority, January 2017), 26.

59. Andrew Thaler, "A Pivotal Moment in the History of Deep-Sea Mining," *DSM Observer*, June 30, 2021, https://dsmobserver.com/2021/06/a-pivotal-moment-in-the-history-of-deep-sea-mining.

60. Daniel Rosenberg, "The Legal Fight over Deep-Sea Resources Enters a New and Uncertain Phase," EJIL:Talk!, August 22, 2023, https://www.ejiltalk.org/the-legal-fight-over-deep-sea-resources-enters-a-new-and-uncertain-phase/.

61. Thaler, "A Pivotal Moment."

62. Thaler.

63. Kaul Gena, "Deep Sea Mining of Submarine Hydrothermal Deposits and Its Possible Environmental Impact in Manus Basin, Papua New Guinea," *Procedia Earth and Planetary Science* 6 (2013): 232.

64. Deep Sea Mining Campaign, "What Is the Precautionary Principle?," http://www.deepseaminingoutofourdepth.org/what-is-the-precautionary-principle/.

65. Reeves, "My Family's Pacific Island Home."

66. Catherine Coumans, "Canada's Role in Deep Seabed Mining," Mining Watch Canada, June 22, 2021, https://miningwatch.ca/blog/2021/6/22/canada-s-role-deep-seabed-mining.

67. Katherine Sammler, "The Deep Pacific: Island Governance and Seabed Mineral Development," in *Island Geographies*, ed. Elaine Stratford (New York: Routledge, 2017), 30.

68. Sammler, 19.

69. Reeves, "My Family's Pacific Island Home."

70. Ben Doherty, "Collapse of PNG Deep-Sea Mining Venture Sparks Calls for Moratorium," Funding the Ocean, September 15, 2019, https://fundingtheocean.org/news/collapse-of-png-deep-sea-mining-venture-sparks-calls-for-moratorium/.

71. Craig Smith, University of Hawai'i, Mānoa, interview with author, September 23, 2019.

72. Dover interview.

73. Kevin Grove and David Chandler, "Introduction: Resilience and the Anthropocene: The Stakes of 'Renaturalising' Politics," *Resilience: International Policies, Practices and Discourses* 5, no. 2 (2017): 81.

74. Melinda Cooper and Jeremy Walker, "Genealogies of Resilience: From Systems Ecology to the Political Economy of Crisis Adaptation," *Security Dialogue* 42, no. 2 (2012): 156.

75. Cooper and Walker, 156.

76. Nautilus Minerals Niugini Limited, *Environmental Impact Statement: Solwara 1 Project*, Volume A Main Report CR 2008_9_v4 (Brisbane, Australia: Coffey Natural Systems Pty Ltd, September 2008), 185.

77. Nautilus Minerals, ch. 9, 21.

78. See Aylin Woodward, "Oil-Eating Microbes Found in the Deepest Part of the Ocean Could Help Clean Up Man-Made Oil Spills," *World Economic Forum*, April 17, 2019, https://www.weforum.org/agenda/2019/04/oil-eating-microbes-found-in-the-deepest-part-of-the-ocean-could-help-clean-up-man-made-oil-spills; Peter Dockrill, "Amazing Plastic-Eating Microbes Could Help Clean Up the Terrible Trash in Our Oceans," *Science Alert*, May 21, 2019, https://www.sciencealert.com/plastic-munching-microbes-in-the-ocean-could-help-fix-our-terrible-pollution-crisis.

79. Although early underwater filmmakers like Jean Painlevé decided to cast shrimp as charismatic protagonists, they did so through a comparative framework that deployed humanity as a measuring stick, rather than simply allowing for radical difference. See James Leo Cahill, *Zoological Surrealism: The Nonhuman Cinema of Jean Painlevé* (Minneapolis: University of Minnesota Press, 2019).

80. Helen Rosenbaum and Francis Grey, "Accountability Zero: A Critique of the Nautilus Minerals Environmental and Social Benchmarking Analysis of the Solwara I Project," Deep Sea Mining Campaign September 29, 2015, 8, http://www.deepseaminingoutofourdepth.org/wp-content/uploads/account abilityZERO_web.pdf.

81. Manon Auguste et al., "Development of an Ecotoxicological Protocol for the Deep-Sea Fauna Using the Hydrothermal Vent Shrimp *Rimicaris exoculate*," *Aquatic Toxicology* 175 (2016): 277–85, https://doi.org/10.1016/j.aquatox.2016.03.024; Chris Hauton et al., "Identifying Toxic Impacts of Metals Potentially Released during Deep-Sea Mining—A Synthesis of the Challenges to Quantifying Risk," *Frontiers in Marine Science* 4, no. 368 (November 16, 2017): 8.

82. Nixon, *Slow Violence*.

83. Cindy Lee Van Dover et al., "Biodiversity Loss from Deep-Sea Mining," *Nature Geoscience* 10, no. 7 (July 2017): 464, https://doi.org/10.1038/ngeo2983.

84. Hessler, *Prospecting Ocean*, 70.

85. Grove and Chandler, "Introduction," 84.

86. Deborah Coen, *Climate in Motion: Science, Empire, and the Problem of Scale* (Chicago: University of Chicago Press, 2018), 3.

87. Anne Pasek, "Mediating Climate, Mediating Scale," *Humanities* 8, no. 159 (2019): 1.

88. Coen, *Climate in Motion*, 4.

89. Orit Halpern, "Hopeful Resilience," *e-flux Architecture*, April 19, 2017, https://www.e-flux.com/architecture/accumulation/96421/hopeful-resilience/.

90. Nassim Nicholas Taleb, *Antifragile: Things That Gain from Disorder* (New York: Random House, 2012), 15.

91. Steve Mentz, *Shipwreck Modernity: Ecologies of Globalization, 1550–1719* (Minneapolis: University of Minnesota Press, 2015), location 105 of 5584, Kindle.

92. Christina Sharpe, *In the Wake: On Blackness and Being* (Durham: Duke University Press, 2016), 18.

93. Astrida Neimanis, "The Sea and the Breathing," *e-flux Architecture*, May 2020, https://www.e-flux.com/architecture/oceans/331869/the-sea-and-the-breathing/.

4. Ocean Pacemakers

1. D. Graham Burnett, *The Sounding of the Whale: Science and Cetaceans in the Twentieth Century* (Chicago: University of Chicago Press, 2012), location 7655 of 20167, Kindle.

2. Arne Kalland, "Management by Totemization: Whale Symbolism and the Anti-whaling Campaign," *Arctic* 46, no. 2 (1993): 125, https://www.jstor.org/stable/40511503.

3. James Bridle, *Ways of Being: Animals, Plants, Machines: The Search for a Planetary Intelligence* (New York: Farrar, Straus and Giroux, 2022), 31.

4. Jennifer Gabrys, *Program Earth: Environmental Sensing Technology and the Making of a Computational Planet* (Minnesota: University of Minnesota Press, 2016), 83.

5. See Lisa Lowe, *The Intimacies of Four Continents* (Durham: Duke University Press, 2015).

6. Hannah Hickey, "Citizen-Science Climate Project Adds Logs from Historic Arctic Whaling Ships," *UW News*, December 3, 2015, https://www.washington.edu/news/2015/12/03/citizen-science-project-adds-whaling-ship-logs-to-study-historic-arctic-climate/.

7. "Killer Whale Up-Close Tour," *Seaworld Orlando*, https://seaworld.com/orlando/tours/killer-whale-up-close-tour/.

8. Kerry Banks, "Tracking Orcas with Tech: 'The Images Took Our Breath Away,'" *The Tyee*, April 15, 2021, https://thetyee.ca/News/2021/04/15/Tracking-Orcas-Tech/.

9. Etienne Benson, *Wired Wilderness: Technologies of Tracking and the Making of Modern Wildlife* (Baltimore: Johns Hopkins University Press, 2010), 23.

10. Mette Bryld and Nina Lykke, *Cosmodolphins* (London: Bloomsbury Publishing, 2000), 14, quoted in Max Ritts and John Shiga, "Military Cetology," *Environmental Humanities* 8, no. 2 (November 2016): 201.

11. Marine Mammal Commission, *Annual Report of the Marine Mammal Commission, Calendar Year 1987* (Washington, D.C.: Marine Mammal Commission, 1988), 156.

12. Bruce Mate, Roderick Mesecar, and Barbara Lagerquist, "The Evolution of Satellite-Monitored Radio Tags for Large Whales: One Laboratory's Experience," *Deep-Sea Research II* 54 (2007): 226–27.

13. Daniel Palacios, interview with author, June 18, 2021.

14. Mate, Mesecar, and Lagerquist, "Evolution of Satellite-Monitored Radio Tags," 225.

15. See chapter 3 of Benson's *Wired Wilderness*, "Diplomatic and Political Subtleties," for more on the ethical debates about tagging.

16. Palacios interview; Rob Harcourt et al., "Animal-Borne Telemetry: An Integral Component of the Ocean Observing Toolkit," *Frontiers in Marine Science* 6, no. 326 (June 2019): 3, https://doi.org/10.3389/fmars.2019.00326.

17. Harcourt et al., "Animal-Borne Telemetry," 3.

18. Palacios interview.

19. Richard Emblin, "Colombia's Pioneer of the Electric Heart," *The City Paper*, November 4, 2015, http://thecitypaperbogota.com/features/colombias -pioneer-of-the-electric-heart/10175.

20. Ana Arana, "Getting to the Hearts of Whales to Assist Human Cardiology," *Baltimore Sun*, October 28, 1990, https://www.baltimoresun.com/news/ bs-xpm-1990-10-28-1990301109-story.html.

21. "Bruce Mate," Oregon State University, https://mmi.oregonstate.edu/ people/bruce-mate.

22. Rob Harcourt et al., "Animal-Borne Telemetry."

23. Gabrys, *Program Earth*, 84.

24. Elizabeth Becker, interview with author, May 6, 2021.

25. "Ocean Currents," *National Geographic*, https://www.nationalgeographic .org/encyclopedia/ocean-currents/.

26. Jessica Lehman, "Making an Anthropocene Ocean: Synoptic Geographies of the International Geophysical Year (1957–1958)," *Annals of the American Association of Geographers* 110, no. 3 (2020): 607, https://doi.org/10.1080/24 694452.2019.1644988.

27. Gabrys, *Program Earth*, 87.

28. Ocean Networks Canada, *Oceans 2.0: An Internet of Things for the Ocean* (Victoria, B.C.: Ocean Networks Canada, 2019).

29. See Yuriko Furuhata, *Climatic Media: Transpacific Experiments in Atmospheric Control* (Durham: Duke University Press, 2022), 111.

30. Jody Berland, *North of Empire: Essays on the Cultural Technologies of Space* (Durham: Duke University Press, 2009), 251.

31. David Lyon and Zygmunt Bauman have spoken of "liquid modernity" and "liquid surveillance" as a way of specifying the immersive qualities of modern surveillance: "Old moorings are loosened as bits of personal data extracted for one purpose are more easily deployed in another. . . . Surveillance spills out all over." Zygmunt Bauman and David Lyon, *Liquid Surveillance: A Conversation* (Cambridge, UK: John Wiley & Sons, 2013), 9.

32. Eugene Thacker, *The Global Genome: Biotechnology, Politics, and Culture* (Cambridge: MIT Press, 2006), 7.

33. Orit Halpern, Robert Mitchell, and Bernard Dionysius Geoghegan, argue that this "optimization fever" universalizes and naturalizes the logic of algorithmic management, "so that optimization's realm can perpetually be expanded and optimization itself further optimized." Orit Halpern, Robert Mitchell, and

Bernard Dionysius Geoghegan, "The Smartness Mandate: Notes toward a Critique," *Grey Room* 68 (Summer 2017): 119.

34. Jessica Lehman, "A Sea of Potential: The Politics of Global Ocean Observations," *Political Geography* 55 (2016): 114.

35. Halpern, Mitchell, and Geoghegan, "The Smartness Mandate," 113.

36. Barbara A. Block et al., "Toward a National Animal Telemetry Network for Aquatic Observations in the United States," *Animal Biotelemetry* 4, no. 6 (2016): 5, https://doi.org/10.1186/s40317-015-0092-1.

37. Block et al., 3.

38. Ana Sequeira, "How Do We Met Future User Needs?," OceanObs'19, Honolulu, Hawai'i, September 16, 2019; Mark Hindell, "Observing the Unobservable: Ocean Observing Data in the Southern Ocean," OceanObs'19, Honolulu, Hawai'i, September 16, 2019.

39. Anne Friedberg, *The Virtual Window: From Alberti to Microsoft* (Cambridge: MIT Press, 2009), 38.

40. Bernhard Siegert, *Cultural Techniques: Grids, Filters, Doors, and Other Articulations of the Real*, trans. Geoffrey Winthrop-Young (New York: Fordham University Press, 2015), 97, 107.

41. Bernard Dionysius Geoghegan, "The Bitmap Is the Territory: How Digital Formats Render Global Positions," *MLN* 136, no. 5 (2021): 1098.

42. Berland, *North of Empire*, 248.

43. Ladd Irvine et al., "An At-Sea Assessment of Argos Location Accuracy for Three Species of Large Whales, and the Effect of Deep-Diving Behavior on Location Error," *Animal Biotelemetry* 8, no 20 (2020), https://doi.org/10.1186/s40317-020-00207-x.

44. Becker interview.

45. The data are, as Lisa Gitelman would say, not raw but "already cooked." Lisa Gitelman, ed., *"Raw Data" Is an Oxymoron* (Cambridge: MIT Press, 2013), 2.

46. Cochran quoted in Benson, *Wired Wilderness*, 30.

47. Hugo Reinert, "Care of Migrants: Telemetry and the Fragile Wild," *Environmental Humanities* 3 (2013): 15.

48. Ritts and Shiga, "Military Cetology," 203.

49. Ritts and Shiga, 207.

50. Raviv Ganchrow, "Earth-Bound Sound: Oscillations of Hearing, Ocean, and Air," *Theory & Event* 24, no. 1 (2021): 94.

51. Lisa Han, "Ichthyoveillance: Fish Spies and the Militarization of Marine Wildlife," *In Media Res*, November 9, 2021, https://mediacommons.org/imr/content/ichthyoveillance-fish-spies-and-militarization-marine-wildlife.

52. Ritts and Shiga, 206.

53. Lauren Holt, "Why the 'Post-Natural' Age Could Be Strange and Beautiful," BBC Future, May 3, 2019, https://www.bbc.com/future/article/201 90502-why-the-post-natural-age-could-be-strange-and-beautiful. For more on posthumanism, see Cary Wolfe, *What Is Posthumanism?* (Minneapolis: University of Minnesota Press, 2009).

54. Block et al., "Toward a National Animal Telemetry Network," 3.

55. NOAA, "What Is an Ocean Gliderr?," National Ocean Service, February 26, 2021, https://oceanservice.noaa.gov/facts/ocean-gliders.html.

56. Michel Foucault, "'Panopticism,' from *Discipline & Punish: the Birth of the Prison*," *Race/Ethnicity: Multidisciplinary Global Contexts* 2, no. 1 (2008): 7.

57. Reinert, "Care of Migrants," 22.

58. Reinert, 9.

59. See Gilles Deleuze, "Postscript on the Societies of Control," *October* 59 (1992): 3–7. http://www.jstor.org/stable/778828.

60. Lehman, "A Sea of Potential," 120.

61. Alexandra Palmer, "The Afterlives of Wildlife Tracking Devices," *Digital Ecologies*, July 6, 2021, http://www.digicologies.com/2021/07/06/alexandra-palmer/.

62. Adam Fish, "Crash Theory: Entrapments of Conservation Drones and Endangered Megafauna," *Science, Technology, & Human Values* 46, no. 2 (2021): 3.

63. Elaina Zachos, "Exclusive: Rare, Mysterious Whales Filmed Professionally for the First Time," *National Geographic*, March 9, 2018, https://www.nationalgeographic.com/animals/article/beaked-whale-drone-photography-first-spd.

64. Carey Kuhn, "Northern Fur Seal Food Study in Bering Sea Using Saildrone: Second Year Adding Video," NOAA Fisheries, *Science Blog*, June 7, 2017, https://www.fisheries.noaa.gov/science-blog/northern-fur-seal-food-study-post-1.

65. Iain Kerr, "SnotBot Goes Tagging: Success!," Ocean Alliance Blog, 2022, https://whale.org/snotbot-goes-tagging-success/.

66. MIT CSAIL, "Robo-Starfish Aims to Enable a Closer Study of Aquatic Life," April 8, 2021, https://www.csail.mit.edu/news/robo-starfish-aims-enable-closer-study-aquatic-life.

67. Mike Fedack, "Animal Platforms: A Cost-Effective Way to Extend the Temporal and Spatial Reach of Other Approaches to Study the Ocean," OceanObs'19, Honolulu, Hawai'i, September 16, 2019.

68. Fish, "Crash Theory," 20.

69. Fish, 6.

70. Palmer, "Afterlives of Wildlife Tracking Devices."

71. Mate et al., "Evolution of Satellite-Monitored Radio Tags," 244–45.

72. Vinicio Lindoso, "Ocean: The Real Lungs of the World," September 13, 2019, UNESCO, https://en.unesco.org/news/ocean-real-lungs-world.

73. Briana Abrahms et al., "Memory and Resource Tracking Drive Blue Whale Migrations," *PNAS* 116, no. 12 (March 2019): 5585.

74. See Wolfe, *What Is Posthumanism?*; Nick Bostrom, "Transhumanist Values," *Journal of Philosophical Research* 30, Supplement (2005): 3–14.

75. Lori Marino et al., "Cetaceans Have Complex Brains for Complex Cognition," *PLOS Biol* 5, no. 5 (2007): e139, https://doi.org/10.1371/journal.pbio.0050139; see also Aviva Hope Rutkin, "Inside the Mind of a Killer Whale: A Q+A with Neuroscientist from 'Blackfish,'" The Raptor Lab, August 14, 2013,

https://theraptorlab.wordpress.com/2013/08/14/inside-the-mind-of-a-killer
-whale-a-qa-with-the-neuroscientist-from-blackfish/.

76. Alexis Pauline Gumbs, *Undrowned: Black Feminist Lessons from Marine Mammals* (Chico, Calif.: AK Press, 2020), 8.

5. Deepwater Feeds

1. Claiborne Pell, "The Scramble Is On for Ocean Riches," *The World*, November 12, 1967, 12.

2. Alfred H. Keil, "Status of Engineering in the Ocean Environment Today," MIT Sea Grant Program, Report No. MITSG 74-6 (November 1, 1973): 1. MIT Oral History Program, oral history interviews on ocean engineering, MC-0089, Box 4, "Transportation by Sea—Today and Tomorrow," by A. A. H. Keil. Department of Distinctive Collections, MIT Libraries, Cambridge, Massachusetts.

3. Paul Fye, "Deep Ocean Technology," unpublished manuscript, July 24, 1964, 4, 9, MC-1, box 3, Project SEABED, Allyn Collins Vine Papers, Woods Hole Data Library and Archives, Woods Hole, Massachusetts.

4. Scott C. Daubin, "Educating and Training Tomorrow's Oceanographers—The Role of Institutions," American Management Association, Washington, D.C., June 20, 1969, 3, MC 0089, box 1, MIT/Woods Hole Oceanographic Institution Ocean Engineering Program, MIT Oral History Program, Massachusetts Institute of Technology, Department of Distinctive Collections, Cambridge, Massachusetts.

5. Alan D. Chave, "Seeding the Seafloor with Observatories: Scientists Extend Their Reach into the Deep with Pioneering Undersea Cable Networks," *Oceanus* 42, no. 2 (April 2004).

6. Helen F. Wilson, "Contact Zones: Multispecies Scholarship through *Imperial Eyes*." *Environment and Planning E: Nature and Space* 2, no. 4 (2019): 720; Mary Louise Pratt, *Imperial Eyes: Travel Writing and Transculturation*, 2nd ed. (London: Routledge, 2008).

7. Arthur Baggeroer et al., "Ocean Observatories: An Engineering Challenge," *The Bridge* 48, no. 3 (September 18, 2018): 19, https://www.nae.edu/195294/Ocean-Observatories-An-Engineering-Challenge.

8. Elizabeth DeLoughrey, "Submarine Futures of the Anthropocene," Oceanic Routes Forum, special issue of *Comparative Literature Journal*, 69, no. 1 (2017): 32–44.

9. Jennifer Gabrys, *Program Earth: Environmental Sensing Technology and the Making of a Computational Planet* (Minneapolis: University of Minnesota Press, 2016), 4.

10. See "Argo Floats," Argo: Part of the Integrated Global Observation Strategy, http://www.argo.ucsd.edu/How_Argo_floats.html.

11. Karen Stocks, "Observing Needs in the Deep Ocean," OceanObs'19 Conference, Honolulu, Hawai'i, September 19, 2019.

12. Eric Lindstrom, "On the Relationship between the Global Ocean Observing System and the Ocean Observatories Initiative," *Oceanography* 31, no. 1 (2018): 40, http://www.jstor.org/stable/26307785.

13. Deep Ocean Observing Strategy, "Background & Need," accessed March 22, 2019, https://deepoceanobserving.org/about/background-need/.

14. Alexander Galloway and Eugene Thacker, *The Exploit: A Theory of Networks* (Minneapolis, University of Minnesota Press, 2007), 81–82.

15. Echoing my mention of "dark mediation" in the introduction, various ocean stakeholders have not only aimed to increase surveillance of unknown ocean spaces but have limited and controlled this audiovisual access.

16. Hitoshi Mikada and Kenichi Asakawa, "Development of Japanese Scientific Cable Technology," *OCEANS* (October 2008): 1–4, https://doi.org/10.1109/OCEANS.2008.5289426.

17. Bruce Howe et al., "Scientific Uses of Submarine Cables: Evolutionary Development Leading to the ALOHA," *Mains'l Haul: A Journal of Pacific Maritime History* 48, nos. 3–4 (Summer/Fall 2012): 107.

18. Broadly speaking, *array* refers to the systematic arrangement of data-collecting instruments. Arrays can exist at multiple scales and can include ocean observatories as well as other complex survey infrastructures, such as the towed hydrophone arrays discussed in chapter 2.

19. Leslie M. Smith et al., "The Ocean Observatories Initiative," *Oceanography* 31, no. 1 (2018): 18.

20. Ocean Networks Canada, "Observatories," 2020, https://www.ocean networks.ca/observatories.

21. Peter Phibbs and Stephen Lentz, "Cabled Ocean Science Observatories as Test Beds for Underwater Technology," *Oceans 2007–Europe* (June 18–21, 2007), https://doi.org/10.1109/OCEANSE.2007.4302268.

22. Phibbs and Lentz, 23.

23. "Cabled Axial Seamount," Ocean Observatories Initiative, 2018, https://oceanobservatories.org/array/cabled-axial-seamount/.

24. Smith et al., "The Ocean Observatories Initiative," 29.

25. "ALOHA Cabled Observatory," University of Hawai'i, last modified June 16, 2022, https://aco-ssds.soest.hawaii.edu/.

26. Nicole Starosielski, *The Undersea Network* (Durham: Duke University Press, 2015).

27. Duncan Geere, "How the First Cable Was Laid across the Atlantic," *Wired*, January 18, 2011, https://www.wired.co.uk/article/transatlantic-cables.

28. Marcie Grabowski, "Deepest Ocean Observatory Celebrates 10 Years of Operation," *University of Hawai'i News*, April 24, 2017.

29. Bruce Howe et al., "ALOHA Cabled Observatory Installation," OCEANS'11—MTS/IEEE Kona, Program Book (2011).

30. Paul Edwards, *A Vast Machine: Computer Models, Climate Data, and the Politics of Global Warming* (Cambridge: MIT Press, 2010), 554.

31. Deborah Kelley, interview with author, Corvallis, Oregon, March 20, 2019.

32. NSF Ocean Observatories Initiative, *OOI Coastal Endurance Array* (Woods Hole, Mass.: Ocean Observatories Initiative, 2018).

33. Kelley interview.

34. Smith et al., "The Ocean Observatories Initiative," 29.

35. Jeff Hecht, "Submarine Cable Goes for the Record: 144,000 Gigabits from Hong Kong to L.A. in 1 Second," *ITU News*, January 5, 2018, https://news.itu.int/submarine-cable-hk-la/.

36. Bruce Howe, "From Space to the Deep Seafloor Using SMART Submarine Cable Systems in the Ocean Observing System," report on two NASA Workshops, September 7, 2015, http://www.soest.hawaii.edu/NASA_SMART_Cables/.

37. There is a precedent for this fusion of sensing and wired transmission in the early history of telephony. Decades before wireless telegraphy was invented, Thomas Watson, Alexander Graham Bell's assistant, first listened to natural radio waves through telephone wires, which were able to sense and transduce electromagnetic currents from the environment: "The sensitivity of the device that made it possible to hear voices also made it possible for Watson to hear natural radio." More than transmitters of intentional signals, telephone wires also acted as sensors, bringing electromagnetic waves into frequencies audible to human hearing. Douglas Kahn, *Earth Sound Earth Signal: Energies and Earth Magnitude in the Arts* (Berkeley: University of California Press, 2013), 27.

38. Alister Hardy, "Was Man More Aquatic in the Past?," *New Scientist*, March 17, 1960, 642–45.

39. Helen Rozwadowski, "Bringing Humanity Full Circle Back into the Sea: *Homo aquaticus*, Evolution, and the Ocean," *Environmental Humanities* 14, no. 1 (March 2022): 23.

40. Jon Crylen, "Living in a World without Sun: Jacques Cousteau, *Homo aquaticus*, and the Dream of Dwelling Undersea," *Journal of Cinema and Media Studies* 58, no. 1 (Fall 2018): 1–23.

41. DeLoughrey, "Submarine Futures of the Anthropocene," 32–33.

42. Eva Hayward, "Sensational Jellyfish: Aquarium Affects and the Matter of Immersion," *differences* 23, no. 3 (2012): 173.

43. These are tactics that Pooja Rangan might call "immediations." Pooja Rangan, *Immediations: The Humanitarian Impulse in Documentary* (Durham: Duke University Press, 2017).

44. Bruce Howe, "A Deep Cabled Observatory: Biology and Physics in the Abyss," *EOS* 95, no. 47 (November 25, 2014): 429–44.

45. Margaret Cohen, *The Underwater Eye: How the Movie Camera Opened the Depths and Unleashed New Realms of Fantasy* (Princeton: Princeton University Press, 2022), 236.

46. Gabrys, *Program Earth*, 57.

47. Nadia Bozak, *The Cinematic Footprint: Lights, Camera, Natural Resources* (New Brunswick, N.J.: Rutgers University Press, 2012), 153.

48. Jim Potemra, interview with author, Honolulu, Hawai'i, September 20, 2019.

49. Lyle Goldstein, "China's 'Undersea Great Wall,'" *The National Interest*, May 16, 2016, https://nationalinterest.org/feature/chinas-undersea-great-wall-16222.

50. Ocean Networks Canada, "Smart Ocean™ Systems," 2020, https://www.oceannetworks.ca/innovation-centre/smart-ocean-systems.

51. H. L. Clark, "New Sea Floor Observatory Networks in Support of Ocean Science Research," Proceedings of the Oceans 2001 MTS/IEEE Conf., Honolulu, Hawai'i (November 5–8, 2001), 5.

52. In chapter 4, I discussed the smartness mandate as it relates to our oceans, and the many projects that work to transform undersea infrastructures into an Internet of Things. Explicitly, undersea cabled observation is connected to smartness through the use of terms such as "Smart Ocean," "SMART cables," and "Smart Oilfields" to describe various proposals for cabled observation platforms.

53. Orit Halpern, Robert Mitchell, and Bernard Dionysius Geoghegan, "The Smartness Mandate: Notes toward a Critique," *Grey Room* 68 (Summer 2017): 114.

54. Howe et al. "Scientific Uses of Submarine Cables," 113.

55. Gabrys, *Program Earth*, 52.

56. Melody Jue, *Wild Blue Media* (Durham: Duke University Press, 2020), 57.

57. Nigel Thrift describes the mathematical mediation of spaces as producing a distinct sense regime of "qualculation," arguing that as locative technology, rendering, and addressability render movement into stable structures for knowledge, "we are increasingly a part of a 'movement-space' which is relative rather than absolute," while relying on "absolute" space as a condition for its existence. Nigel Thrift, "Movement-Space: The Changing Domain of Thinking Resulting from the Development of New Kinds of Spatial Awareness." *Economy and Society* 33, no. 4 (2004): 597, https://doi.org/10.1080/0308514042000285305.

58. Paolo Favali, Laura Beranzoli, and Angelo De Santis, *Seafloor Observatories: A New Vision of the Earth from the Abyss* (Chichester: Springer Praxis Books, 2015), 5–6.

59. Ann Elias, *Coral Empire: Underwater Oceans, Colonial Tropics, Visual Modernity* (Durham: Duke University Press, 2019), 51.

60. Ocean Networks Canada, *Oceans 2.0: An Internet of Things for the Ocean* (Victoria, B.C.: Ocean Networks Canada, 2019).

61. David Aubin, Charlotte Bigg, and H. Otto Sibum, eds., *The Heavens on Earth: Observatories and Astronomy in Nineteenth-Century Science and Culture* (Durham: Duke University Press, 2010), 29.

62. Ludwik Fleck, "Schauen, sehen, wissen," 154, quoted in Aubin, Bigg, and Sibum, `19.

63. Aubin, Bigg, and Sibum, 31.

64. Mark D. Wilkinson et al., "The FAIR Guiding Principles for Scientific Data Management and Stewardship," *Nature*, March 2016, 1, 3, https://www.nature.com/articles/sdata201618.

65. Favali, Beranzoli, and de Santis, *Seafloor Observatories*, 131.

66. Kelley interview.

67. Smith et al., "The Ocean Observatories Initiative," 33.

68. Brooks A. Kaiser et al., "The Importance of Connected Ocean Monitoring Knowledge Systems and Communities," *Frontiers of Marine Science* 6 (June 14, 2019), https://www.frontiersin.org/articles/10.3389/fmars.2019.00309/full.

69. See "The Aha Moku System," https://www.ahamoku.org/.

70. Paulokaleioku Timmy Bailey, "Kupaianaha Indigenous Event," Ocean-Obs'19 Conference, Honolulu, Hawai'i, September 16, 2019.

71. Shelly Denny, "Integrating Western and Indigenous Knowledge Systems: Two-Eyed Seeing in Nova Scotia," OceanObs'19 Conference, Honolulu, Hawai'i, September 19, 2019.

72. Halpern, Mitchell, and Geoghegan, "The Smartness Mandate," 122–23.

73. Denny, "Integrating Western and Indigenous Knowledge Systems."

74. Joshua Schuster, "Coral Cultures in the Anthropocene," *Cultural Studies Review* 25, no. 1 (2019): 91, https://doi.org/10.5130/csr.v25i1.6405.

75. Anna Tsing, *Friction: An Ethnography of Global Connection* (Princeton: Princeton University Press, 2005), 75.

76. Ursula Heise, *Sense of Place and Sense of Planet: The Environmental Imagination of the Global* (Oxford: Oxford University Press, 2008), 54.

77. Heise, 56.

78. Tsing, *Friction*, 59.

79. Karen Amimoto Ingersoll, *Waves of Knowing: A Seascape Epistemology* (Durham: Duke University Press, 2016), 549.

80. Marisa Elena Duarte, *Network Sovereignty: Building the Internet across Indian Country* (Seattle: University of Washington Press, 2017), 6.

81. For more on indigenous storywork, see Jo-ann Archibald Q'um Q'um Xiiem, Jenny Bol Jun Lee-Morgan, and Jason De Santolo, *Decolonizing Research: Indigenous Storywork as Methodology* (New York: Bloomsbury, 2019).

82. Potemra interview.

83. Smith et al., "The Ocean Observatories Initiative," 30–31.

84. Potemra interview.

85. Smith et al., "The Ocean Observatories Initiative," 31.

86. The caption text goes on to explain, "The build-up of marine life on instruments—known as biofouling—can interfere with their ability to gather data. While this can present a significant problem for ocean observing, it's also important to appreciate the beauty of these diverse underwater communities."

87. See Richard Maxwell and Toby Miller, *Greening the Media* (New York: Oxford University Press, 2012).

88. Karen Barad, "Invertebrate Visions: Diffractions of the Brittlestar," in *The Multispecies Salon*, ed. Eben Kirksey (Durham: Duke University Press, 2014), 227.

Conclusion

1. Alfred A. H. Keil, "The New Challenge of Ocean Engineering," *Technology Review* 69, no. 6 (March 1967): 5, MIT Oral History Program, oral history interviews on ocean engineering, MC-0089, box 4, "Transportation by Sea— Today and Tomorrow," by A. A. H. Keil, Department of Distinctive Collections, MIT Libraries, Cambridge, Massachusetts.

2. Keil, 1.

3. Alfred Keil, "The Development of Ocean-Engineering (An Attempt at a Methodology)," Department of Naval Architecture and Marine Engineering (September 8, 1966): 8, MIT Oral History Program, oral history interviews on ocean engineering, MC-0089, box 4, "Transportation by Sea—Today and Tomorrow," by A. A. H. Keil, Department of Distinctive Collections, MIT Libraries, Cambridge, Massachusetts.

4. Lesley Green, "Oceanic Regime Shift," in *Sustaining Seas: Oceanic Space and the Politics of Care*, ed. Elspeth Probyn, Kate Johnson, and Nancy Lee (New York: Rowman and Littlefield, 2020), 15–17.

5. Stuart Hall, "The Spectacle of the 'Other,'" in *Representation: Cultural Representations and Signifying Practices*, ed. Stuart Hall (London: Sage, 1997), 261.

6. Helen Rozwadowski, "Bringing Humanity Full Circle Back into the Sea: *Homo aquaticus*, Evolution, and the Ocean," *Environmental Humanities* 14, no. 1 (March 2022): 16.

7. Macarena Gómez-Barris. *The Extractive Zone: Social Ecologies and Decolonial Perspectives* (Durham: Duke University Press, 2017).

8. Shelly Denny, "Integrating Western and Indigenous Knowledge Systems: Two-Eyed Seeing in Nova Scotia," OceanObs'19 Conference, Honolulu, Hawai'i, September 19, 2019.

9. Paul Gilroy, "'Where every breeze speaks of courage and liberty': Offshore Humanism and Marine Xenology, or, Racism and the Problem of Critique at Sea Level," The 2015 Antipode RGS-IBG Lecture, *Antipode* 50, no. 1 (2018): 17.

10. James Bridle, *Ways of Being: Animals, Plants, Machines: The Search for a Planetary Intelligence* (New York: Farrar, Straus and Giroux, 2022), 297.

11. Malreddy Pavan Kumar, "(An)other Way of Being Human: 'Indigenous' Alternatives to Postcolonial Humanism," *Third World Quarterly* 32, no. 9 (2011): 1568.

12. Sebastian De Line, "All My/Our Relations: Can Posthumanism Be Decolonized?," July 7, 2016, Open! Platform for Art, Culture & the Public Domain, http://onlineopen.org/all-my-our-relations.

13. Astrida Neimanis, "The Sea and the Breathing," in *Sustaining Seas: Oceanic Space and the Politics of Care*, ed. Elspeth Probyn, Kate Johnston, and Nancy Lee (New York: Rowman & Littlefield, 2020), 209.

14. Sarah Dimick, *Unseasonable: Climate Arrhythmias in Global Literature* (New York: Columbia University Press, forthcoming).

15. Bernard Dionysius Geoghegan, *Code: From Information Theory to French Theory* (Durham: Duke University Press, 2023).

16. I borrow the term "oppositional gaze" from bell hooks, who uses it to describe a critical gaze back at power; an oppositional gaze endows agency to those who are typically subordinated to a controlling and hegemonic gaze, as was the case for black female spectators watching cinema that failed to represent them. bell hooks, "The Oppositional Gaze: Black Female Spectators," in *Black Looks: Race and Representation* (Boston: South End Press, 1992), 115–31.

17. Susan Schuppli, *Material Witness: Media, Forensics, Evidence* (Cambridge: MIT Press, 1994).

18. Schuppli, 3.

19. Schuppli, 4.

20. Stacy Alaimo, "Violet-Black," in *Prismatic Ecology: Ecotheory beyond Green*, ed. Jeffrey Jerome Cohen (Minneapolis: University of Minnesota Press, 2013), 235.

21. Sofia Crespo, "What Can Exploring the Digital Representation of Aquatic Life Teach Us?" (Presentation, Ocean Memory Project Senses and Sensing Seed Seminar, Zoom, February 10, 2022).

22. Donna Haraway, "Staying with the Trouble for Multispecies Environmental Justice," *Dialogues in Human Geography* 8, no. 1 (2018): 102–5.

23. *Plastified: An Ode*, dir. Alice Aires (Royal College of Art, 2020), https://www.aliceaires.net/plastified-an-ode.

INDEX

Abrahms, Briana, 151
absent bodies, 52–53
accountability, 14, 109, 113
acoustic mediation, 60
activism, 59, 92, 107–13
adaptation, 12, 114–16, 118, 166, 193
affect, 1, 15, 76, 79, 88, 108, 142; and nonhuman animals, 85–86, 123, 125–26, 152
affordances, 50–51, 96–97, 126, 174, 191
afternoon effect, 64
agency, 14, 19, 31, 97, 138, 140, 194; human, 8, 40, 54–55, 125, 174, 185; nonhuman animals', 31, 49–51, 186, 191, 196; ocean, 94, 118, 159, 180
Aires, Alice, 199–200
air gun arrays, 22, 61, 71–72, 78, 82, 85–86
Alaimo, Stacy, 10, 199
Alberti's velo, 139
alchemy, 4–6, 14, 16, 25
Alford, Matthew, 103, 105
ALOHA Cabled Observatory (ACO), 163–65, 168, 170–71, 176–77, 181
Amazon River, 129, 132
American frontier, 43, 46–47
American Petroleum Institute (API), 73, 86–87
ancient civilizations, 31, 40

animal-borne sensors (ABS), 143–44, 149
animal studies, 81
Anthropocene, the, 32, 42, 87, 93; and seabed mining, 101–2, 105, 110, 120; and seabed observatories, 158–59, 174
anthropocentrism, 133, 185, 188, 193, 195–96; and ocean mediation, 14, 78, 125, 140
anthropogenic disturbances, 103, 116, 132, 198
anthropogenic sound, 78, 81, 84, 86, 89
anthropology, 32, 38, 59, 77, 123, 167, 191
anthropomorphism, 137, 174
anti-blackness, 120
antifragility, 118–19
anxiety, 7, 33, 64, 85, 93, 96, 107
aqua homo, 168, 174, 180
Aquaman (film), 185
aquariums, 35, 84, 101, 127, 136, 168, 172
aquatic entertainment industry, 129
archaeological excavation, 29, 34–40, 49–52
archaeology, 32–33, 35–39, 40, 47, 50
archival tags, 125, 131, 134
archives: archivability, 39, 138; archival mediation, 35–36, 40, 197;

LISA YIN HAN is assistant professor of media studies at Pitzer College.

Printed in the USA
CPSIA information can be obtained
at www.ICGtesting.com
CBHW081220260824
13213CB00020B/21